THE PHOENIX PROJECT
Shifting from Oil to Hydrogen

By

HARRY BRAUN

Revised Second Edition
First Printing: 2000

Published in 2000 in the United States of America by:

SPI PUBLICATIONS & PRODUCTIONS
6245 North 24th Parkway, Suite 208
Phoenix, Arizona, 85016
Telephone: (602) 957-1818
Fax: (602) 957-1919
www.phoenixproject.net

LIBRARY OF CONGRESS
CATALOGING IN PUBLICATION DATA:
Braun, Harry W., III, 1948-
The Phoenix Project: Shifting form Oil to Hydrogen
1. Energy. 2. Environment. 3. Hydrogen. 4. Biotechnology

Library of Congress Number: 89-60277

ISBN: 0-9702502-0-7 (Hard Cover)

Printed and bound in the U.S.A.

To Sir Isaac Newton,
for his remarkable insights
and his dedication to critical thinking.

FOREWORD

The Phoenix Project is an analysis and synthesis of some of the most significant information and ideas that could allow the U.S., in cooperation with other countries, to make an energy and industrial transition to renewable energy resources and technologies. If this transition is successfully completed, it will insure that the advances in molecular biology will result in a biological transition to renewable resources. If the transition is not undertaken before energy and agricultural systems fail, it may not happen at all. In this regard, The Phoenix Project is a comprehensive review of both the problems and the promise that confront the global human community.

There are many books and articles that have described the global environmental problems related to greenhouse gases, acid rain, stratospheric ozone depletion, and chemical contamination. The Phoenix Project is unique because it also defines a specific and well-documented plan to fundamentally resolve these problems. The essence of the solution involves an energy transition from the existing "oil economy" to a "hydrogen economy." This transition to a hydrogen energy system is being advocated by thousands of scientists and engineers from over 80 countries, but thus far, most Members of the Congress, the media and the general public are simply unaware of the issue.

The Phoenix Project is a code name similar to the "Manhattan Project" in World War II, except that instead of concentrating the talent of the best and brightest technical minds on developing weapon systems, they would instead be developing renewable energy and biological technologies. While some people will assume that the oil companies will never allow such a transition to occur, many of the major oil companies are at the forefront of the transition. Indeed, oil companies will evolve into hydrogen production and distribution companies because it will be in the economic self-interest of their stockholders to do so.

Rather than drilling deep, dry holes in sensitive offshore areas or the remaining wilderness areas to find the last of the oil, energy companies need to redirect their investments into the relatively low-risk renewable energy technologies that can make the U.S. energy independent and essentially pollution-free.

Unlike oil and natural gas wells that eventually run dry, the hydrogen energy system is inexhaustible. As a result, the transition to renewable resources will be a win-win situation for both industry and environmentalists. It is my hope that *The Phoenix Project* will help to serve as a trigger mechanism for change, because humanity is as close to a kind of technological utopia as it is to oblivion. We are living in the midst of an unprecedented information explosion, and given the limited amount of time that each person has to absorb and assimilate data, one of the most important questions to ask is: *What is worth knowing?* This question underscores the necessity of sifting through the sea of data to extract the significant insights. There is not enough time to know everything, which now means that one of the most important educational functions is to prioritize information.

The Phoenix Project is a proposed plan that has evolved over many years of research. It is a synthesis of key points and insights that have been published in scientific and technical journals, newspapers, news magazines, books, or broadcasts on network news over a period of many years. The individual bits of information are very much like the pieces to a puzzle, in that after a threshold of pieces have been put in place, one can begin to see "the big picture." Although many technical issues are reviewed, no prior technical or mathematical skills are required to understand any of the key points presented.

Perhaps most importantly, *The Phoenix Project* documents that it is possible to have "prosperity without pollution," and that the primary obstacles to implementing such a transition are not technical, but political. It has been said that Caesar found Rome as a city of clay and left it a city of marble. This generation, which inherited a highly polluted civilization based on non-renewable fossil and nuclear fuels, now has the opportunity -- *and the responsibility* -- to leave our descendants with an inexhaustible energy source that will make prosperity without pollution a reality.

--HB

ACKNOWLEDGMENTS

I am especially indebted to the following individuals for their many special contributions and insights that have made this book possible. As has been said: *"If we are able to see farther than some, it is only because we are standing on the shoulders of many who have gone before us."* The following individuals, who are listed alphabetically, are certainly in that category.

Albert Bartlett, Ph.D., Department of Physics, University of Colorado (Boulder, Colorado), for his many efforts to communicate the significance of exponential growth. I am also grateful to Renz Jennings, the former Chairman of the Arizona Corporation Commission) for making me aware of Dr. Bartlett's work.

Glendon Benson, Ph.D., President, Aker Industries (Oakland, California), for his extensive engineering research and development of advanced solar Stirling engine generator-sets (gensets) for hydrogen production.

Roger Billings, Director, American Association for Science (Independence, Missouri), for his life-long research and development of hydrogen energy production and end-use systems.

Jim Blackmon, Ph.D., Director of Advanced Programs for Boeing, for his many insights and contributions to the development of concentrating solar power systems.

Christer Bratt, the Stirling engine program manager for Kockums Submarine Systems, for his many contributions to the development of Stirling engines.

G. Daniel Brewer, P.E., Lockheed Corporation (Burbank, California), for the many contributions that he and his colleagues at Lockheed have made in the development of liquid hydrogen-fueled aircraft and related support systems.

David Dressler and Huntington Potter, for their remarkable and enlightening book: *Discovering Enzymes.*

K. Eric Drexler, for his remarkable book, *Engines of Creation,* that provides such a clear vision of nanotechnology and how it will provide the basis for a sustainable utopia.

William E. Heronemus, for his many thoughtful insights regarding wind and ocean thermal energy conversion systems.

Caesar Marchetti, Ph.D., International Institute for Applied Systems Analysis (Austria), for his perspective and many insights on how the Earth's initial microbial "founding fathers" developed the first hydrogen energy system over three billion years ago.

Hans Nelving, for his many thoughtful insights and contributions to the development of Stirling engines and electrical power conversion units.

Gerard O'Neill, Ph.D., for his book, *The High Frontier: Human Colonies in Space,* which provides a realistic engineering and economic analysis of large-scale space colonization.

Robert L. Pons, P.E., engineering program manager for Ford Aerospace & Communications Corporation, for his many insights into the similarities between manufacturing solar point-focus-concentrator systems and automobiles.

Harry Reid, for his many years of support in the U.S. Senate for renewable energy and hydrogen energy systems.

Stanley Schneider, Senior Technical Fellow with Boeing, for his efforts to significantly expand Boeing's role in the development and commercialization of renewable hydrogen energy technologies.

Richard (Dick) Shaltens, NASA Glenn Chief engineer, Thermo Mechanical Branch, for his commitment to the optimization of Stirling engines and related subsystems.

David Slawson, President, Stirling Energy Systems (Phoenix, Arizona), for his commitment to the commercialization of renewable energy technologies.

Walter F. Stewart, and his colleagues at the Los Alamos National Laboratory, for their photographs and many insights into the use and handling of liquid hydrogen.

Ken Stone, and his many colleagues at Boeing who have done so much with so little to advance the solar Stirling and advanced wind technologies.

Wolfgang Strobl, who BMW refers to as the "father of the BMW liquid hydrogen-fueled car program," and his many colleagues at BMW for their unshaken and genuinely inspiring determination to demonstrate that liquid hydrogen-fueled cars are indeed practical.

Lubert Stryer, at Stanford University for his remarkable book: *Molecular Design of Life.*

T. Nejat Veziroglu, Ph.D., and his colleagues who make up the International Association for Hydrogen Energy. Their extraordinary efforts in creating and managing an international peer-review hydrogen technical society have provided the foundation upon which an industrial transition to renewable resources can be built. I am also deeply indebted to the other scientists and engineers that have prepared the papers that serve as the technical basis of this book. These papers have been published by *The American Institute of Aeronautics and Astronautics, American Journal of Physics, Chemtech, IEEE Spectrum, The International Journal of Hydrogen Energy, Mechanical Engineering, Nature, Oceanus, Science, Scientific American, and the Society of Automotive Engineers.*

I am especially grateful to Dorothy L. Hays for her tireless devotion and her extraordinary organizational skills, as well as Pete Dixon and John Olson, without whose assistance the second edition of this book would not have been possible. I am also indebted to Jay O'Malley, former Chairman of the Board of the O'Malley Companies (Phoenix, Arizona) who, in spite of the objections of many, allowed myself and our chief engineer, Tom Bird, the opportunity to conduct the extensive research that serves as the foundation of this book. Finally, I wish to thank my sister Kathy and her husband Gary for their patience and understanding over the years, and my parents who were always there when I needed them.

-- HB

TABLE OF CONTENTS

Chapter 1: Utopia or Oblivion   1
The Problem   1
The Hydrogen Economy   4
Resolvable Environmental Problems   7
Global Acid Deposition   7
Greenhouse Gases   9
The Carbon Cycles   11
Stratospheric Ozone Depletion   14
Photobiology   22
A Question of Balance   25
Chemical Contamination   27
Projections   31
Trends   33
Conclusions   35

Chapter 2: Exponential Icebergs   37
Albert A. Bartlett   39
The Concept of 11:59   42
The Age of Exponentials   44
Positive Exponentials   46
The Information Explosion   47
Future Shock   48
Strategies for Survival   49
Education: Problems and Solutions   50
Starting Young   51
Teaching Teachers   52
Multiple Exponential Problems   54
Conclusions   55

Chapter 3: Fossil Fuels   57
Interrelationships   57
Energy Fundamentals   61
Coal   62
Crude Oil Reserves   63
Exponential Expiration Time   65
Petroleum Reserve Estimates   67
The Battery   68
Natural Gas   70
Conclusions   71

Chapter 4: Nuclear Power   73
Nuclear Fission   73
Radioactivity   75
Decommissioning   76
Nuclear Waste Storage   78
Thermal Load   80
Questions of Safety   83
Advanced Reactor Considerations   84
Nuclear Fusion   86
Cold Fusion   86
Nuclear Economics   87
Conclusions   90

Chapter 5: Hydrogen   93
In the Beginning   93
Primordial Hydrogen   96
The Nanobes & Microbes   97
Photosynthesis   102
The Water-Former   104
The Universal Fuel   105
Primary vs. Secondary Energy Sources   107
Hydrogen Storage Systems   108
Hydrogen Hydrides   109
Liquid Hydrogen   113
The Hydrogen Engine   119
Roger Billings   120
Hydrogen Safety   124
Hydrogen Explosions   127
Hydrogen Aircraft Applications   129
NASA   136

Hypersonic Aerospacecraft   139
Space Habitats   142
Hydrogen Starships   144
Space Colonies   145
Other Alternative Fuels   149
Hydrogen Production   151
Conclusions   153

Chapter 6: Renewable Energy Technologies   159
Energy Economics   160
External Costs   161
Photovoltaic Cells   162
Solar Engine Systems   164
Early Solar Stirling Systems   166
Stirling-Cycle Engines   174
Manufacturing Costs   186
BioStirling Systems   187
Thermoacoustic Stirling Engines   188
Central Receiver Systems   190
Line-Focus Systems   192
Comparative Analysis   194
Wind Energy Conversion Systems   195
Dispatchability   200
Siting   202
Offshore Wind Systems   203
Vertical Vortex Wind Systems   207
OTEC   211
Economic Considerations   220
U.S. Energy Policy   222
Conclusions   225

Chapter 7: Renewable Energy Resources   229
Solar Resources   230
Wind Energy Resources   239
OTEC Resources   240
Implementation Lead-Times   241
Water Considerations   243
Water Options   245
NAWAPA   247
OTEC Sea Water Desalination   251
Conclusions   252

Chapter 8: Utopia: From Here to Eternity   255
    Designer Genes   256
    Blue Gene   258
    The Biochip   260
    Immortalizing Enzyme   262
    Molecules of Memory   264
    Biocybernetic Evolution   266
    The Creators   269
    Nanobial Origins   272
    Nanobes Evolve Into Microbes   273
    Microbial Memory   279
    The City of the Cell   287
    The Human Genome Project   290
    Nanocomputers   292
    Microtechnology   294
    Nanoscopes   296
    Resonance-Ionization Spectroscopy   298
    Overpopulation Considerations   299
    The Cryogenic Ark   301
    Defining Death   303
    The Ultimate Disease   305
    Ethical Questions   307
    Priorities   309
    Summary   310

Chapter 9: Priorities   313
    Solutions   314
    Lifeboat Agricultural Systems   315
    Political Priorities   319
    Government Regulation   321
    Fair Accounting Act   323
    The Phoenix Project   332

References   341
    Chapter 1   341
    Chapter 2   344
    Chapter 3   344
    Chapter 4   345
    Chapter 5   346
    Chapter 6   349

Chapter 7   351
Chapter 8   351
Chapter 9   355

Index   357

The Author   366

THE PHOENIX PROJECT

Figure 1.1: The Earth

UTOPIA OR OBLIVION

The Problem

In exploring the vast universe, astronomers and other space scientists have found that there are more than a hundred billion galaxies. Within each galaxy are hundreds of billions of stars, and orbiting most stars is a substantial collection of planets, moons and asteroids. This means there are theoretically many billions of planets that are similar to the Earth. At present, however, the Earth is the only planet that is known to be capable of sustaining human life. In spite of this remarkable fact, the surface of the Earth may soon become uninhabitable for humans and other mammals. This unfortunate event will not likely occur as a result of a nuclear war, but as a result of the exponentially increasing chemical contamination of the Earth's soil, water and atmosphere, along with the physical destruction of its ancient biological habitats. Such environmental damage is somewhat like a nuclear war occurring in slow motion.

The needless death of the only planetary life support system known to support mammalian life by a seemingly mindless "bulldozer culture" is no small consideration. It means that the era of mammals, which has been evolving for over 100 million years, may soon be coming to an end. For many species, it is already too late. Indeed, there are now as many species being exterminated as during the great mass extinction that ended the era of the dinosaurs. This time, however, it is not the reptiles that are on the threshold of extinction, but the mammals. The question is: *Are there going to be any survivors?*

There is an extraordinary irony that given the exponential advances in information, particularly in the areas of molecular biology and computer science, it is reasonable to assume that disease as it is now understood, as well as other molecular disorders could soon be eliminated. In the coming era of "designer genes," such molecular-scale technologies could also be directed to repair the Earth's environmental life-support systems that are presently being destroyed and/or contaminated on a global scale. Such developments, however, are predicated on the survival of an urban-industrial civilization that allows tens of thousands of highly trained scientists to concentrate in highly specialized fields. The problem is that at this point, not only is the survival of civilization at risk, but the survival of the species. The details and implications of the exponential nature of the global problems will be discussed in more detail in Chapter 2, but in the final analysis, it appears that the human community is simultaneously racing toward both oblivion and utopia.

It is the central thesis of this book that although the oblivion scenario is likely, it is neither necessary nor inevitable. Given the extensive global ecological damage that has already occurred, it is highly likely that regardless of what actions are taken now, a significant price is going to be paid for the reckless disregard of the Earth's biological life support systems. However, if it is known that there are major problems ahead, it is possible to be prepared and thereby minimize the damage. Moreover, the major obstacles to change are not technical, but political. The solution essentially involves the industrialized world making a transition to renewable energy and biological resources. But the most important consideration is that there is only a limited opportunity to take the corrective action necessary because the global environmental problems are deteriorating exponentially.

Although the global environmental problems are in large part related to energy and industrial policies, there is no question that the problems are being compounded by the exponential growth in the human population and its resulting increased consumption of the Earth's nonrenewable resources. A major concern is the fact that primary environmental, energy and economic problems that have been accumulating gradually over many years are not responsive to sudden changes. These global forces are like an automobile that is increasing its speed as it heads for a cliff. There is a point-of-no-return beyond which brakes and steering are useless. Once that threshold is passed, it will not matter if the driver suddenly applies the brakes with all his strength and attempts to alter course. Even with all the wheels locked, the force of momentum will propel the automobile over the cliff into an uncontrollable disaster.

Because a large ship has a much greater mass and momentum than an automobile, it requires much longer lead-times to change course. Countries are much larger than ships, and global atmospheric and biological life-support systems are much larger than countries, which means the lead-times to take corrective action are much longer still. This is an important consideration whether one is dealing with the accumulation of pesticides in food chains, the accumulation of greenhouse gases in the atmosphere, or the depletion of the Earth's stratospheric ozone layer. This is the nature of the problems we face.

It is tragic that the destruction of the Earth's biological life-support systems does not need to occur in order for urban-industrial populations to maintain their automobiles and high-technology life-styles. Assuming it is not too late to take corrective actions, it is helpful to know what could make a difference in minimizing the impact of the many global environmental disruptions that now appear inevitable.

One thing is clear. Many of the most serious global environmental and economic problems are related to the fact that the industrialized world is based on a fossil fuel energy system that does not factor-in the environmental costs that are inevitable with such a system. Fossil fuels have a negative environmental impact from the point of exploration, recovery, transportation of the fuel, and in their final end-use when they are burned in the atmosphere. To resolve the problem, it will be necessary for the industrialized world to make a transition to a non-fossil fuel energy system that is capable of sustaining an increasing human urban-industrial population. Such an energy option does exist. It has been referred to as the hydrogen energy system, but given the more comprehensive implications that are involved, many distinguished engineers and scientists simply refer to the new energy system as the "hydrogen economy." This book will document a specific proposal and plan that will allow the U.S., in cooperation with other countries worldwide, to make an energy, industrial and biological transition to renewable solar-hydrogen resources by 2010 or 2020.

Such a transition could ensure global economic progress while allowing the industrialized world to begin living in harmony and balance with the Earth's biological life support systems. "Phoenix" is an ancient Egyptian word that refers to a mythical bird that would never die. After having lived for several hundred years, this immortal bird would then consume itself in flames, and rise out of its own ashes to be reborn. In a similar context, it is technically possible for the industrialized world to rise from the ashes of the highly polluting fossil fuels (principally oil, coal and natural gas), and renew itself by developing renewable solar-hydrogen technologies and resources that are both inexhaustible and essentially pollution-free.

The Hydrogen Economy

Hydrogen is not just another energy option like nuclear, solar, petroleum or coal. Rather, it is a "universal fuel" that can unite virtually all energy sources with all energy uses. Hydrogen can be converted into electricity, and electricity can be turned into hydrogen. Hydrogen is the most abundant element in the known universe. It is the primary fuel for the Sun and other stars, and it can be manufactured from a wide range of sources, such as natural gas, biomass resources or water. When hydrogen is burned as a combustion fuel, the resulting by-product is water vapor, making it completely renewable.

Hydrogen is essentially pollution-free when burned and is generally safer than gasoline in the event of an accident or collision. The transition to a hydrogen economy is also one of the key variables that makes an industrial transition to renewable solar resources technically and economically feasible. For this reason, Chapter 5 is devoted to hydrogen and its origin, production, and potential use as an energy medium. There already is an extensive scientific and engineering brain trust, the International Association for Hydrogen Energy (Coral Gables, Florida), which has over 2,500 representatives from over 80 countries. These distinguished scientists and engineers hold international technical conferences every two years to review virtually every aspect of hydrogen production, transportation, storage and end-use applications. Organizing and coordinating a fundamental industrial transition to renewable solar-hydrogen resources, however, is an enormous task somewhat comparable to the Manhattan Project or setting up the U.S. War Production Board in World War II. As a result, such a fundamental energy transition cannot happen without widespread public and political support.

There are many precedents for such an effort. During World War II the U.S. War Department organized the "Manhattan Project," which involved bringing together a large number of key scientists and engineers to focus upon the development of the atomic bomb. The American space program, the development of the railroads, or the interstate highway system are other examples of long-term "macro-engineering" projects that could not have occurred without the successful cooperation of the government and private industry. A similar type of focused effort will be required if the human community is to accomplish a successful transition to a renewable resource "stable state" energy and economic system -- before the Earth's biological life support systems are irreparably damaged. After a certain point, it may be impossible to implement such a transition because the Earth's natural resources are being exponentially depleted at the same time that its biological life support systems are being exponentially destroyed and/or contaminated.

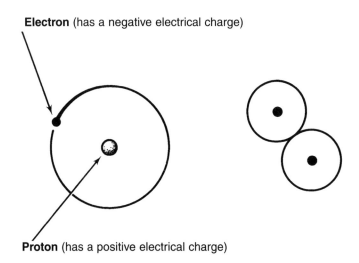

Electron (has a negative electrical charge)

Proton (has a positive electrical charge)

Figure 1.2: The Hydrogen atom (H₁) & molecule (H₂)

There already are a number of reasons to be concerned, but perhaps the single most ominous development is the depletion of the Earth's stratospheric ozone layer that for several billion years has shielded the biological life on Earth from the high energy and extremely deadly short-wave cosmic radiation. The loss of the stratospheric ozone shield will result in the collapse of agricultural systems worldwide, and the result of such a development would obviously be catastrophic.

Civilization is a highly complex network of interdependent forces, but when major food-production and the related economic systems begin to disintegrate, social stability can disappear in a matter of weeks or days -- if not hours. This is particularly true when one considers the psychological impact that could result when large numbers of people who are totally dependent on supermarkets, discover that they may soon be running out of food. Imagine the run on the local supermarkets by panic-stricken people who quickly realize that they have very little food stored. The stores would be emptied of food and other basic goods in a matter of hours. Most people have enough food to last several days, but if assistance is not forthcoming, or expected, the outlook for the future would be grim.

There is already roughly one-quarter of the human population that is without adequate food and fresh water, but in a global systems collapse the vast majority of the Earth's human population would be lost in the chaos that would soon follow. If things are allowed to progress to this stage, an industrial transition to renewable resources will be as meaningless as trying to cash a check in a supermarket empty of food. The oblivion scenario will have come to pass.

In order to cope with a potential collapse of existing global food production systems, it is necessary to be prepared for the multiple environmental dislocations that are, to a certain extent, now inevitable. By acknowledging that there are many serious problems ahead and by planning accordingly, it is possible to minimize the impact of such problems. It will hopefully be possible to avoid the complete breakdown of the Earth's existing biological life support systems, but at present, no one can know for sure. While the global energy and related environmental problems are formidable, however, there are also many options to deal with the problems that will be documented in the subsequent chapters of this book.

Resolvable Environmental Problems

As an initial step in planning for constructive changes in the future, it is helpful to have a basic understanding of the global environmental problems that could be significantly resolved by making a transition to renewable solar-hydrogen energy technologies and resources. Some of the most serious problems that could be fundamentally resolved by making such a transition include the following:

Global acid deposition: Acid deposition includes wet acid rain, fog and snow, as well as dry acid particles that corrode buildings and bridges, and destroy biological habitats.

Greenhouse gases: Greenhouse gases that result in global warming and climate change are principally made up of carbon dioxide, methane, and chlorofluorocarbon (CFC) molecules.

Stratospheric ozone depletion: Chlorine-based chemicals, principally CFCs, as well as gases such as methane are chemically dissolving the Earth's protective ozone layer.

Chemical contamination: Benzene emissions, oil spills, leaks from gasoline and diesel fuel storage tanks, and the production of both high level and low level radioactive wastes.

Global Acid Deposition

Acid rain is the term that has received the most use in the media, but acid deposition (i.e., fallout) refers to a mixture of both wet and dry compounds of sulfur and nitrous oxides. The wet part of acid deposition can be in the form of rain, snow, hail, sleet, dew, frost or fog. A solution's acidity is measured on the pH scale by its concentration of hydrogen ions. Because pH value is a logarithm of the reciprocal of the hydrogen ion concentrations, pH falls as acidity rises, and each full unit of change on the pH scale represents a tenfold increase or decrease in acidity.

The acidity range is from 1 (highly acidic) to 14 (highly alka-
line). A value of 7 is neutral. A lake becomes acidified when its
pH falls below 6. Water collected near the base of clouds in the
eastern U.S. during the summer now has a pH of about 3.6, but
values as low as 2.6 have been recorded. The effects of acid
deposition on forests have been devastating. According to a
paper that appeared in the August 1988 issue of *Scientific
American* by Volker A. Mohnen, a professor of atmospheric science at the State University of New York at Albany:

> *"Since 1980 many forests in the eastern U.S. and
> parts of Europe have suffered a drastic loss of
> vitality -- a loss that could not be linked to any of
> the familiar causes, such as insects, disease or
> direct poisoning by a specific air or water pollu-
> tant. The most dramatic reports have come from
> Germany, where scientists, stunned by the
> extent and speed of the decline, have called it
> Waldsterben, or forest death. Statistics for the
> U.S. are also unnerving."*

Acid fog can be even worse than acid rain. Scientists at the
California Institute of Technology and the University of California
at Los Angeles have found acid fog to be 100 times more acidic
than acid rain. In addition, the dry acid particles and gases are
also a serious problem. According to Robert W. Shaw, Chief of
Chemical Diagnostics and surface science at the U.S. Army
Research Office, the dry deposition can be as destructive as the
"wet" compounds. In a paper published in the August 1987 issue
of *Scientific American*, Shaw indicated that acid deposition is
generated primarily by the burning of fossil fuels:

> *"Studies tracing particle samples to their sources
> have helped to quash the notion that natural
> emissions from swamps, volcanoes or trees
> might be responsible for the acid fallout around
> the globe. It is now beyond question that even in
> rural areas acid deposition (both wet and dry)
> almost always stems from the activity of human
> beings: primarily the combustion of fuel for
> power, industry and transportation."*

In excess of 40 million tons of acid deposition is emitted into the atmosphere annually just from the U.S. Acid deposition not only severely affects biological systems, it also corrodes buildings, bridges, and other structures as well as the priceless ancient statues in Europe and the Middle East that have survived for centuries. While the vast majority of insects and microorganisms have been able to adapt to the pesticides that have thus far been developed, and even moderate exposure to radioactive isotopes -- acid deposition is another matter. This is because the acids dissolve the very biochemical molecules that make life possible in the first place. Acid deposition has not received as much media coverage as the global warming greenhouse gases, but it is every bit as serious a concern.

Greenhouse Gases

Greenhouse gases, which are principally made up of carbon dioxide, methane and the chlorofluorocarbons, trap the infrared heat from the Sun just like the glass in a greenhouse, thereby raising atmospheric temperatures. One of the many ways this "greenhouse effect" can directly impact weather conditions is by increasing droughts, which, in turn, contributes to the advancing of the deserts. Globally, it has been estimated by Richard A. Houghton and George M. Woodwell, two senior scientists at the Woods Hole Research Center in Woods Hole, Massachusetts, that the net atmospheric gain of carbon dioxide is about 3 billion tons annually.

Houghton and Woodwell indicated in their April 1989 issue of *Scientific American* article, "Global Climate Change," that while the total amount of carbon dioxide in the atmosphere is only about three one-hundredths of one percent, in contrast to oxygen and nitrogen that together make up 99 percent of the atmosphere, oxygen and nitrogen are not greenhouse gases that absorb infrared radiation or radiant heat, whereas the greenhouse gases like carbon dioxide, methane and chlorofluorocarbons do. This means that in spite of their relatively small concentrations, the greenhouse gases play a significant role in regulating the overall temperature of the Earth. Woodwell has also provided congressional testimony that the widespread destruction of forests is significantly accelerating the global warming trend because the dying forests release, as carbon dioxide, the carbon stored in their organic matter. In addition, Syukuro Manabe and Richard Wetherald, of the U.S. National Oceanic and Atmospheric Administration's Fluid Dynamics Laboratory at Princeton University, have reported that carbon dioxide increases have also been implicated in drying out the soil in major agricultural areas of the U.S. Their study, published in the April 1986 issue of *Science,* indicates that the rising carbon dioxide levels could trigger a significant reduction in soil moisture in the grain belts of the U.S., Canada and Europe.

James Hansen, Director of the National Aeronautics and Space Administration's (NASA) Institute for Space Studies and an expert on climate change, testified before the Senate Energy and Natural Resources Committee on June 23, 1988, that the Earth had been warmer in the first five months of 1988 than in any comparable period since measurements began 130 years ago. Hanson attributed the increase in temperature to the long expected greenhouse effect, stating that "...it is time to stop waffling so much and say that the evidence is pretty strong that the greenhouse effect is here."

While predicting climate change is an inherently complex process, the vast majority of atmospheric scientists agree with Jerry Mahlman, Director of the National Oceanic and Atmospheric Administration's Geophysical Fluid Dynamics Laboratory at Princeton University, who believes that global warming is not only well underway, but it's increasingly obvious that industrial output of greenhouse gases -- principally carbon dioxide -- is responsible for it. Computer driven climatological models that have been run by Mahlman and colleagues have been matching forecasts with actual conditions that have developed over the past decade. These data have confirmed Mahlman's observations; and in his view, the only debate left has to do with the extent of the climate changes that are now inevitable. In December of 1998 in Kyoto, Japan, 159 nations agreed to reduce carbon dioxide emissions to below 1990 levels by 2012, but even with such cutbacks, carbon dioxide levels will triple by 2100. Thus, the best Kyoto can do is produce a small decrease in the rate of increase. This problem is compounded by the fact that the existing political leadership in the Congress refuses to ratify even the existing Kyoto Protocols, much less the more significant carbon dioxide reductions that are necessary.

The Carbon Cycles

According to Wallace Broecker, a professor of geochemistry at Columbia University and Director of the Center for Climate Research, the thermostat that determines the surface temperature of the Earth is the amount of carbon dioxide that is in the atmosphere, which in turn, is determined by the carbon cycle. For hundreds of millions of years, the carbon cycle was maintained by rain-washing the carbon dioxide from the atmosphere into the oceans, where the concentration is 50 times that of the atmosphere. The carbon dioxide is gradually deposited on the ocean floor, where it combines with calcium to form limestone. The limestone then moves with the geologic plates until it is returned to the molten interior of the Earth. Eventually, the carbon is returned to the atmosphere as carbon dioxide through volcanic eruptions. This geologic process results in an overall carbon cycle time of about 200 million years.

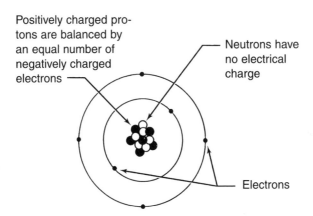

Positively charged protons are balanced by an equal number of negatively charged electrons

Neutrons have no electrical charge

Electrons

Only the electrons in the outer energy level are involved in chemical reactions

Figure 1.3: The Carbon Atom
The carbon atom has six protons and six neutrons in its nucleus. Surrounding the nucleus in orbital paths are six electrons. Two of the electrons are located in the inner energy level, and the remaining four chemically active electrons are in the outer energy level, or valence.

Another major carbon cycle that has a relatively rapid cycle time involves living organisms. All life on the Earth is based on the metabolic assimilation of carbon and hydrogen. The protein molecules within green plants, for example, extract carbon dioxide out of the air, and by using the energy from the Sun to also extract hydrogen from water. Utilizing a process referred to as photosynthesis, they are able to combine the carbon and hydrogen to form energy-rich carbohydrates (i.e. molecules of hydrogen and carbon). A by-product of this reaction is oxygen, which is released into the atmosphere. Indeed, the oxygen that was used to create the stratospheric ozone shield was generated by this photosynthetic process.

When the plant is consumed by other living organisms, energy is extracted from the carbohydrates, and the waste product of this cellular metabolism is carbon dioxide, which is again returned to the atmosphere. If the plant material is not consumed and decomposes into one of the fossil fuels, the carbon is stored until such time as the fossil fuels are burned. Ice cores that have been cut from the polar ice caps have allowed scientists to analyze air that was trapped in the ice for thousands of years. As a result, it has been determined that at the peak of the Ice Age, some 20,000 years ago, the amount of carbon dioxide in the atmosphere was about 30 percent less than it is today.

Since the geologic carbon cycle takes hundreds of millions of years, Broecker believes that the relatively rapid carbon dioxide change may have been brought about by changes in the ocean population of plankton. These microscopic plants which are members of the vast and ancient algae family, anchor the aquatic food chain. Plankton metabolize significant amounts of carbon dioxide, and when they die, they sink to the ocean floor, carrying the carbon with them. Thus, the greater the population of plankton, the more carbon dioxide they are able to extract from of the atmosphere. When the plankton population declined during the last Ice Age, more carbon dioxide was retained in the atmosphere. This caused the temperature of the Earth to increase, which may have brought an end to the last Ice Age.

A more fundamental explanation has been put forth by James E. Lovelock, a distinguished atmospheric chemist in the U.K. who believes the sum total of the Earth's microbial life forms, including the vast and ancient populations of bacteria, fungi and molds, make up a global living system.

Lovelock refers to this global biological "superorganism" as Gaia, a Greek word that means "Goddess of the Earth." He believes that this Gaian system has been actively regulating the climate of the Earth for over 3 billion years by engineering life forms that generate or absorb various atmospheric gases, including carbon dioxide, oxygen and methane. Lovelock summarized his theory in his books, *Gaia: A New Look at Life on Earth,* and *The Ages of Gaia: A Biography of Our Living Earth.* He points out that although the Sun has increased its heat output by roughly 30 percent since life began on the Earth, the surface temperature has remained relatively constant.

Lovelock believes this highly unnatural phenomenon may have occurred because the sum total of the Earth's microbes, which were concentrated in the continental shelves of the oceans and forests on land, were able to pump just enough carbon dioxide out of the atmosphere to keep the surface of the Earth from heating up along with its aging sun. However, Lovelock believes that the Gaian force is reaching the end of its ability to regulate the Earth's climate. This is because the Sun is continuing to increase its heat output at the same time that carbon dioxide and other greenhouse gases are increasing due to the combustion of fossil fuels and the simultaneous destruction of the remaining rain forests.

The logical conclusion is that if the Gaian forces are no longer capable of removing carbon dioxide from the Earth's atmosphere, the existing climatic "stable-state" equilibrium is soon going to be lost. This means the human population and other mammalian life forms may soon be extinguished in the rapidly changing climatic environment -- in the same way that the dinosaurs disappeared within a relatively brief period of time. The major groups of microbes will surely survive for a time by engineering new organisms that will be able to adapt to the more severe environment. However, the surface of the Earth will be a vastly different place than it is now.

Regardless of whether a microbial superorganism exists or not, it is clear that the delicate balance of carbon dioxide and other gases in the Earth's atmosphere is a critical factor in allowing living organisms to survive and evolve. From the beginning of the industrial revolution, increasing amounts of carbon dioxide have been released into the atmosphere, principally as a result of deforestation and the combustion of fossil fuels.

In the U.S., which has the highest per capita consumption of fossil fuels of any country, about six tons of carbon dioxide are released per person annually. Global carbon dioxide levels have increased by roughly 10 percent since 1900, and the rate is continuing to increase. Venus provides a graphic example of what happens when too much carbon dioxide accumulates in the atmosphere. Venus has no water to wash the carbon dioxide from its atmosphere, thus it has a surface temperature of about 850 degrees Fahrenheit. Mars, on the other hand, is an example of what happens when there is too little atmospheric carbon dioxide. Initially, Mars had oceans and a climate similar to that of the Earth. But because the continents did not drift on Mars, the carbon was trapped on the ocean floor. This resulted in increasingly cooler surface temperatures until the oceans finally froze.

Stratospheric Ozone Depletion

Of all of the environmental problems, none is more serious than the depletion of the stratospheric ozone layer, which has protected life on the surface of the Earth for approximately 3 billion years from the deadly high-energy (below 290 nm) solar radiation. The stratospheric ozone shield was initially generated as a result of photosynthetic microorganisms extracting hydrogen from water, and releasing oxygen to the atmosphere as a by-product. As the oxygen drifted up into the Earth's upper atmosphere, it was converted by ultraviolet energy into ozone, which is made up of three atoms of oxygen. This ozone is continuously being created and destroyed by the Sun's ultraviolet radiation, but up until recently, it has remained in balance.

What has changed is that hundreds of thousands of tons of chlorine-based chlorofluorocarbon (CFC) molecules have been pumped into the atmosphere over the last five decades. CFCs are used as industrial solvents and in a wide range of products, including aerosol spray cans and foam packaging. But some of the most ozone-reactive CFCs are those that have been used in the Freon™ of air conditioning systems. CFCs account for about 20 percent of the greenhouse gases, and also contribute to the acid rain problem, but their most serious effect is that they are one of the principal agents responsible for destroying the Earth's stratospheric ozone layer.

As the CFCs drifted up into the upper atmosphere, they would eventually chemically break down from exposure to the ultraviolet radiation, thereby releasing the chlorine. Once released, a single chlorine atom can react with and destroy over 100,000 stratospheric ozone molecules. Solar radiation is made up of electromagnetic units that Albert Einstein referred to as photons. The energy level of a photon is dependent on its wavelength (refer to Figures 1.4 and 1.5); the shorter the wavelength, the higher the energy level. It is significant that as the wavelength decreases, the energy level of the photon increases exponentially. Biological organisms are made up of molecules that are, in turn, made up of a highly complex arrangement of atoms. Each atom contains a nucleus made up of protons (hydrogen nuclei) and/or neutrons, which are usually surrounded by a cloud of one or more orbiting electrons. Electrons carry a negative electrical charge, and they orbit the positively-charged nucleus somewhat like planets orbiting a star. As the electrons absorb photons of various energy wavelengths, they eventually become sufficiently energized to change their energy level, or break away from the nucleus of the atom altogether. This movement of electrons is the very basis of molecular chemistry, which includes everything from the synthesis of deoxyribonucleic acid (DNA) in the nucleus of our cells to making a cup of coffee.

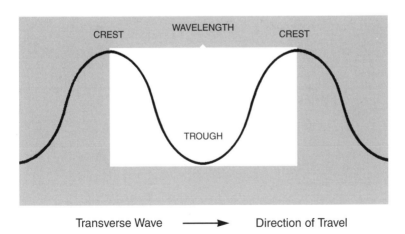

Figure 1.4: Wavelength

In the case of the high-energy short-wave ultraviolet radiation, the energy of the photon is so great that it can easily destroy the delicate three-dimensional electrochemical structure of biological molecules. Such molecular damage can cause healthy cells to mutate into cancer cells, or to die altogether. Up until recently, the Earth's protective ozone shield had been absorbing all of the dangerous ultraviolet photons (i.e., those with a wavelength that is below 290 nanometers). However, as CFC molecules began to be manufactured in larger and larger quantities in the 1950s and 1960s, they began to unravel one of the critical elements of the Earth's life-support system. By the 1980s, roughly 800,000 tons of CFCs were being released into the Earth's atmosphere annually. It was also in the early 1980s that scientists began to record serious depletions or "holes" in the Earth's stratospheric ozone layer. The serious nature of stratospheric ozone depletion cannot be overstated. Life cannot survive on the surface of the Earth without it! If the Earth's protective ozone layer continues to be degraded, agricultural systems worldwide will collapse, and the urban chaos that will then be inevitable could rapidly lead to the oblivion scenario.

Electric Waves: 3,100 miles
(= 4,988 Kilometers)

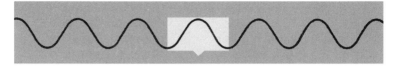

Ultraviolet Radiation: 300 nanometers
(Note: A nanometer is one-billionth of a meter)

Figure 1.5: Wavelength and Energy Level
As the wavelength decreases, its energy level increases exponentially.

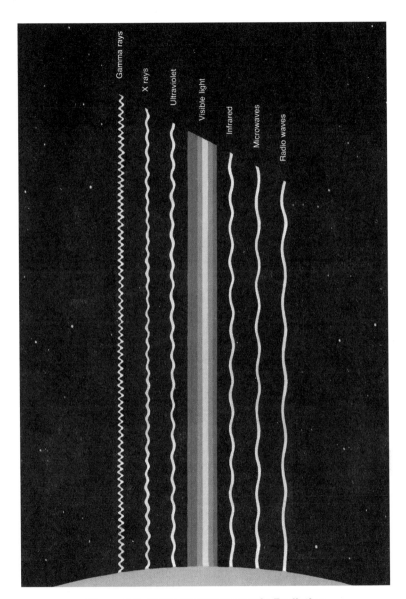

Figure 1.6: Solar Electromagnetic Radiation
Solar radiation from our Sun is emitted in a broad range of energy levels. For several billion years, the Earth's stratospheric ozone shield has been able to filter out the deadly shortwave electromagnetic radiation below 290 nanometers. With "holes" in the stratospheric ozone layer that are as large as the continental U.S., our critical shield is being lost.

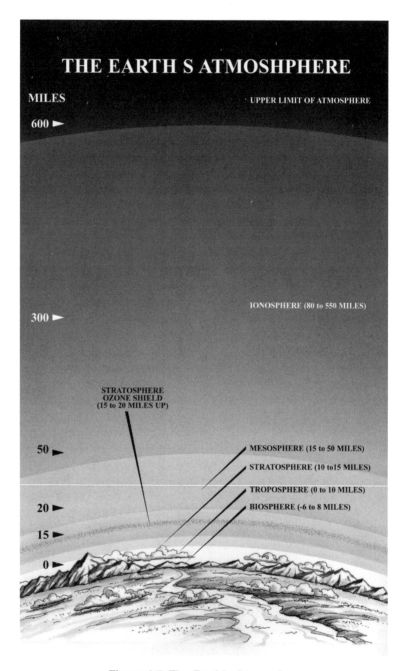

Figure 1.7: The Earth's Atmosphere

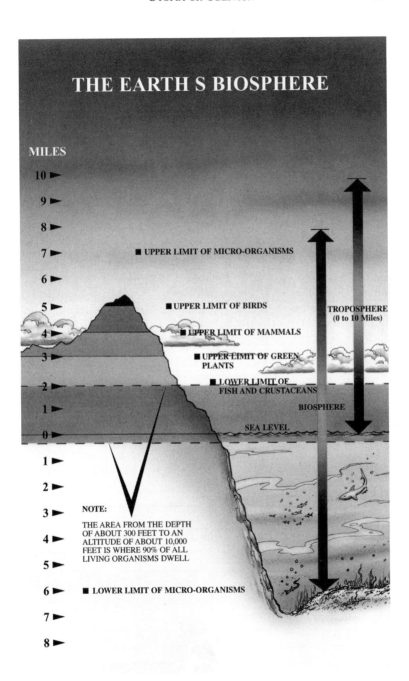

Figure 1.8: The Earth's Biosphere

The first major hole in the stratospheric ozone layer was discovered in 1981 by a small British research team, the British Antarctic Survey, using a single, relatively unsophisticated instrument. At first, the British researchers, under the direction of J. C. Farman, thought the instrument they were using had been giving them false readings. They were aware that NASA and other scientific teams had been using highly sophisticated instrumentation to measure stratospheric ozone levels, and they could not figure out why these other laboratories had not announced such a significant drop in the Antarctic ozone levels.

What they did not know was that the NASA scientists had programmed their high-speed supercomputers to ignore any ozone readings that were off by more than a few percentage points, assuming them to be in error. Up to that point, most scientists believed that drops of 20 or 30 percent in the stratospheric ozone shield was not even possible. The measurements of Farman and his colleagues however, indicated that the Antarctic ozone levels had decreased by an alarming 50 percent just within the last decade. After carefully rechecking his data, Farman and his British colleagues published their results in 1985. Farman's initial results were essentially ignored by the media until the scientists at NASA went back and reexamined their computer data (which comes out at the rate of 200,000 data points per day) and confirmed Farman's observation that the massive stratospheric ozone hole was indeed there. The hole was the size of the continental U.S. and could be seen from Mars! In addition to the large stratospheric ozone hole in the Antarctic areas, overall global stratospheric ozone readings during the past ten years have also indicated an overall decline of approximately three to four percent.

There is, unfortunately, another more ominous kind of instrumentation that is confirming the fears regarding stratospheric ozone depletion -- skin cancer. According to Darrel Rigel, a research physician from the New York University Medical Center in New York, the rate of skin cancer in the U.S. is already increasing at a near epidemic pace -- outstripping predictions initially made in the 1980s. In testimony before a U.S. Senate Hearing of the Energy and Commerce Health and Environmental Subcommittee on Ozone Depletion, Rigel stated that malignant melanoma, which is one of the deadliest forms of skin cancer, increased 83 percent from 1980 through 1987.

Melanoma is now increasing faster than any other form of cancer in humans with the exception of lung cancer in women. According to Rigel:

> *"Five years ago it was unusual to see people under 40 with skin cancer. Now, we often find it in people in their 20s."*

Robert Watson, director for upper-atmospheric research at NASA, has made similar observations:

> *"There is now compelling observational evidence that the chemical composition of the atmosphere is changing at a rapid rate on a global scale."*

Although the potential damage of CFCs to the Earth's ozone shield was widely reported in the press in the mid-1970s, the CFC industries continued to disregard such warnings as "alarmist," which allowed them to maintain and even expand their multi-billion dollar business. In 1977, the Environmental Protection Agency (EPA) banned the use of CFCs in aerosol spray cans, but the CFC industries found other applications for their chemicals. As a result, the worldwide production of CFCs actually continued to increase at about 5 percent per year. An international treaty to reduce the amount of CFCs by 50 percent by the year 2010 was signed in 1988 by the major CFC producing countries. But most atmospheric scientists believed that a 50 percent reduction was inadequate. An international treaty was finally signed in 1996 to ban the further use of ozone depleting chemicals, but the latest reports from NASA regarding stratospheric ozone depletion are not encouraging. As of April 2000, the Earth's protective ozone layer has thinned to record low levels around the Arctic region. Moreover, the thinning ozone layer may be headed for even more dramatic losses because of global warming. The fact that exposure to the natural outdoor sunlight is becoming increasingly risky has ominous biological implications for the public at large because extensive scientific literature has documented that exposure to the natural outdoor sunlight can be as important to human health and productivity as nutrition and exercise.

Photobiology

Given that the energy levels of electromagnetic radiation (EMR) exponentially increase as the wavelength decreases, being exposed to solar radiation that is below 290 nanometers (nm) is clearly a dangerous activity. However, what is not well known is that extensive clinical and laboratory data have also shown that *a lack* of exposure to 290 nm photons of EMR can also routinely induce profound pathological effects in humans and other animals. For millions of years the EMR that has reached the surface of the Earth has been made up of ultraviolet (UV) and visible spectrum that is generally referred to as sunlight. As the illustrations in Figures 1.9 and 1.10 shows, 290-nanometer wavelength UV radiation has a surprisingly wide impact on human health and productivity.

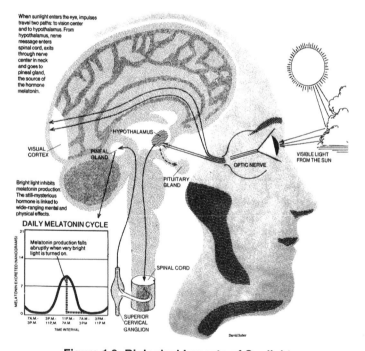

Figure 1.9: Biological Impacts of Sunlight
High-energy 290 nm wavelength photons initiate the release of key hormones that regulate the body's immune and endocrine system. Illustration provided courtesy of the *New York Times*.

Some of the direct and indirect effects of light on the human body are outlined in Figure 1.10. The biological pathways for light include the eyes, which are connected to the brain and neuroendocrine organs as well as the skin, where 290 nm ultraviolet radiation is needed to manufacture Vitamin D, which allows the body to assimilate calcium and other minerals from food. Without Vitamin D, the body will rob the calcium from the teeth and bones.

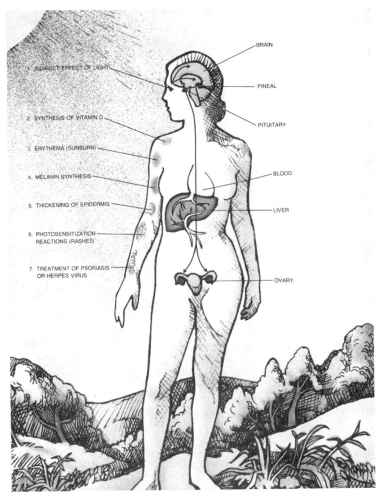

Figure 1.10 Sunlight and Health
The illustration was provided courtesy of *Scientific American*.

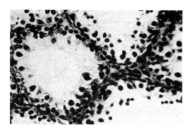

Figure 1.11
The testis from a hamster raised under cool white lamps were malformed and had no sperm production.

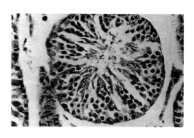

Figure 1.12
The testis from a hamster raised under full spectrum lamps were normal and had normal sperm production.

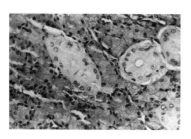

Figure 1.13
The submandibular gland tissue from a hamster that was raised under cool white lamps was malformed.

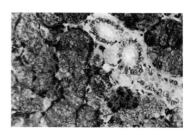

Figure 1.14
The submandibular gland tissue from a hamster raised under full spectrum lamps was normal in appearance and function.

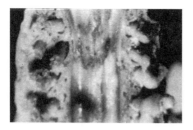

Figure 1.15
The molar teeth from a hamster raised under cool white lamps had 500% more tooth decay that was 10 times larger than normal.

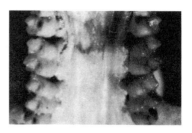

Figure 1.16
The molar teeth from the hamsters that were raised under full spectrum fluorescent lamps had normal tooth decay.

A Question of Balance

The news media have provided the general public with an abundance of articles that discuss the risks associated with over-exposure to sunlight, which can cause skin cancer -- and even death. What has not been clarified is that exposure to EMR is similar to exposure to water. If one gets too much water at any given time, one drowns. On the other hand, water in moderate, daily amounts is critical for human health. So it is with EMR. While some dermatologists argue that *any* exposure to ultraviolet wavelengths is dangerous, a wide-range of critical metabolic functions, such as Vitamin D formation absolutely requires the highly energetic 290 nm ultraviolet wavelength photons. If individuals (or the golden hamsters that are pictured in Figure 1.15) are deprived of exposure to ultraviolet radiation, their bodies leach the calcium from the teeth and bones.

The quality -- as opposed to quantity (i.e. brightness) -- of light is something that is rarely given serious thought by most people. As a result, it is rarely understood, even by many medical professionals, that visible light is only a small part of the EMR spectrum that was a primal factor in the evolution of life itself. The EMR that actually reaches the surface of the Earth is made up of both visible and invisible ultraviolet components of the spectrum. The common term is sunshine, but the technical term is solar global radiation (SGR), which refers to both the direct and indirect (scattered Sun and sky) sunlight that has been specifically defined in terms of its nanometer (nm) wavelength. SGR is fairly stable in the near ultraviolet component of the spectrum (320 to 380 nm) and visible (380 to 750 nm) spectrum, while the middle ultraviolet wavelengths (290 to 320 nm) vary with the angle of the Sun.

Ordinary window glass reflects or absorbs much of the biologically active EMR, and since the 1950s, architects and engineers have been designing buildings with windows that cannot be opened -- or without windows at all. Thus millions of urban inhabitants have been inadvertently sealed-off from the most energetic spectra of terrestrial solar EMR. The only EMR that is provided in the interior spaces of "modern" buildings (which have been referred to as "sick buildings") comes from the fluorescent lighting system whose EMR is typically substantially different than that which occurs naturally in the outdoor environment.

Although "full spectrum" fluorescent lamps that simulate the natural outdoor EMR have been commercially available in the U.S. since the 1960s, they are currently not used by any major company, hospital chain or school district. This issue has come to the attention of a few distinguished scientists, including Dr. Richard J. Wurtman, who is professor of endocrinology and metabolism at the Massachusetts Institute of Technology. Wurtman also lectures at the Harvard Medical School where he received his M.D. degree in 1960. In his landmark paper, "The Effects of Light on the Human Body," published in *Scientific American* in July 1975, Wurtman made the following observation:

". . . only minimal sums have been expended to characterize and exploit the biological effects of light, and very little has been done to protect citizens against potentially harmful or biologically inadequate lighting environments. Both government and industry have been satisfied to allow people who buy electric lamps . . . to serve as the unwitting subjects in a long-term experiment on the effects of artificial lighting on human health."

If the 60 golden hamsters that participated in a photobiology study by investigators at Harvard University's School of Dental Medicine are any indication, there are reasons to be concerned. In this well-controlled study that eliminated environmental variables of diet and exercise, the only difference was the types of fluorescent lamps that the animals were exposed to. Half of the hamsters were housed under full spectrum lamps that simulated the natural outdoor environment, and the other half were placed under the "cool white" lamps that are in widespread use. After 15 weeks of exposure the animals were sacrificed and examined. The animals in the cool white environments had 500 percent more tooth decay, and the caries were 10 times as large as the animals that were raised under the full spectrum lamps. Moreover, the animals raised under the cool white lamps had smaller bodies, hearts and sex organs, and the males had no sperm production. Other significant histological differences were also observed in the submandibular gland structures. While additional data is needed, it is clear that while too much exposure to solar radiation is dangerous, too little exposure will also result in a wide-range of medical disorders.

Chemical Contamination

Photochemical smog is a principal component of urban air pollution worldwide, and it typically includes dangerous increases in tropospheric ozone (i.e., that found in the lower atmosphere that extends from the Earth's surface to about 10 miles out). According to two Cornell University scientists, Peter Reich and Robert Amundson, tropospheric ozone can cause even more short-term damage to plants than acid rain. Tropospheric ozone increases are to be distinguished from the stratospheric ozone levels that are being dramatically decreased.

Other common, yet serious environmental contaminants include the dumping of toxic crankcase oil that is used to lubricate the tens of millions of automotive engines. There is also the unfortunate reality that there are literally millions of gallons of gasoline that leak into the ground annually from the storage tanks of neighborhood gas stations and other hydrocarbon fuel storage facilities. According to EPA, there are an estimated 75,000 to 100,000 leaking fuel storage tanks that are releasing roughly 11 million gallons of gasoline annually into the ground, and the number of faulty tanks is increasing as the corrosion process continues with time. Officials from the American Petroleum Institute have confirmed that the petroleum industry's estimates are about the same as the EPA estimates.

Among the toxic substances in gasoline are benzene and ethylene dibromide, both of which are known to cause cancer. Because of the toxicity of gasoline, the EPA scientists indicated that such pollution poses a serious threat to the nation's underground water supplies. Jack Ravin, an EPA administrator for water pollution, testified before a Senate hearing that one gallon of gasoline a day leaking into an underground water source is enough to contaminate the water supply of a community of 50,000 people. That means that if 11 million gallons are leaking into underground water supplies annually, over 150 million people in the U.S. could be affected by these hazardous compounds. Thus even so-called "solutions" to the air pollution problems can generate significant environmental problems of their own. Methyl tertiary butyl ether (MTBE), for example, was developed by oil companies in the 1970s in order to boost octane and replace the lead in gasoline. Now it is known that MTBE is one of the most insidious forms of water pollution.

Figure 1.17: Oil Spills
Most of the oil that is spilled in the oceans does not occur from the highly publicized accidents but from routine tanker operations.

**Figure 1.18
Offshore Accidents**
Accidents are an unavoidable part of any human activity, including energy production, transportation and distribution. In the case of accidents that involve oil and other hydrocarbon fuel spills, the damage to the natural environment can last for many decades.

Figure 1.19: Air Pollution
Urban air pollution is a common problem worldwide. Detailed studies have shown that the vast majority of air pollution in urban areas is caused by the use of gasoline and diesel fuel in the transportation sector. Even with the vastly improved emissions from new automobiles, the submicron-sized carbon particles still continue to be emitted, which results in the urban haze that has become a characteristic of modern urban life.

Figure 1.20:
Nuclear Power Plants
Nuclear power plants do not emit visible air pollution or carbon emissions that can impact climate change, but they have other significant environmental problems that include the long-term storage of the radioactive wastes.

MTBE is referred to as an "oxygenated" fuel because it adds oxygen to gasoline, which improves combustion efficiency and thereby reduces tailpipe emissions from 20 to 40 percent. Following the passage of the Clean Air Act in 1990, federal authorities in the U.S. ordered the phase-in of such oxygenated fuels in the nation's smoggiest urban areas. By 1999, over 4.5 *billion* gallons of MTBE was being manufactured annually to be blended with the gasoline that is used in roughly 70 percent of the nation's automobiles. However, both federal and state officials are seeking to ban the continued use of MTBE because it has leaked from underground gasoline storage tanks, and in the process, it has contaminated drinking water and groundwater supplies throughout the U.S.

A European study has linked MTBE to liver and kidney tumors in mice, although the long-term health effects on humans are unknown. MTBE is a particularly potent pollutant because it has a very high chemical solubility, which allows it to spread very rapidly in water. Dennis Cocking of the South Tahoe Public Utility District, where 12 of 34 wells were closed because of MTBE, described MTBE to an Associated Press reporter John Howard as " a diabolical chemical" because it moves so rapidly through his community water systems. Doug Marsano of the Denver-based American Water Works Association, a consortium of water agencies, has urged President Clinton to ban MTBE. Marsano points out that MTBE spreads exponentially:

> *"Once you find out you have a problem,*
> *you have a big problem.*
> *And once it's in, how do you get it out?"*

According to Marsano, the chemical has now been detected in varying amounts in all 50 states. Significant MTBE contamination has been found in such pastoral towns as Ronan, Montana as well as in virtually all major urban areas. In Santa Monica, California, MTBE contamination has shut down half of its wells and it is believed the cleanup could cost over $100 million. Even in its tiniest proportions of only five parts per billion, MTBE, which smells like turpentine, has an easily detectable smell. Although it is not known if humans and other mammals will be able to adapt to such air and water pollution, the current cancer statistics are not encouraging.

According to the American Cancer Society, more than a million U.S. citizens discover they have cancer annually, bringing the total to over 86 million. Roughly 500,000 Americans die from cancer each year, making it the fourth leading cause of death. Moreover, there are other serious global problems that compound the chemical contamination problems. Among these is a loss of forests (deforestation). From a historical perspective, the process of human civilization has been based on the consumption of trees. Over 90 percent of the natural forests of the Earth are gone, and as humans continue to consume the remaining forests, they eliminate the very trees that remove carbon dioxide from the air, as well as attract clouds and the life-dependent moisture. Deforestation accelerates the erosion of topsoil and thereby contributes to the advancing of deserts. Desert regions, in turn, reflect the Sun's heat creating a high-pressure system that makes it that much more difficult for the cooler moist air to enter these areas. This deadly cycle feeds on itself, and the net result can be devastating to agricultural systems.

Projections

Because of the complexity of global atmospheric and weather systems, no one can accurately predict how soon these problems will begin to affect the Earth's major food production systems. However, given the exponential nature of these problems, a major collapse of the Earth's life support systems could occur within the next 30 to 50 years, if not sooner. Indeed, serious food and water shortages are already occurring in many parts of the world. In a study funded by the U.S. Department of Energy, over 100 million weather records from 1861 to 1984 were analyzed to determine if any climate trends could be identified. The study not only concluded that the planet was growing warmer, but that the trend is accelerating. The hottest three years of the 123-year period were 1980, 1981, and 1983, and five of the nine hottest years occurred after 1978. These data, coupled with the fact that a new record heat wave occurred again in 1988, prompted many atmospheric scientists to speculate that the dreaded greenhouse warming is beginning to affect global weather conditions. In their *Scientific American* article, "Global Climatic Change," Houghton and Woodwell summarized their findings as follows:

"The world is warming. Climatic zones are shifting. Glaciers are melting. Sea levels are rising. These are not hypothetical events from a science fiction movie -- these changes and others are already taking place, and we expect them to accelerate over the next years as the amounts of carbon dioxide, methane and other trace gases accumulating in the atmosphere through human activities increase. The warming, rapid now, may become even more rapid as a result of the warming itself, and it will continue into the indefinite future unless we take deliberate steps to slow or stop it. Those steps are large and apparently difficult: a 50 percent reduction in the global consumption of fossil fuels, a halting of deforestation, a massive program of reforestation. There is little choice. A rapid and continuous warming will not only be destructive to agriculture but also lead to the widespread death of forest trees, uncertainty in water supplies, and flooding of coastal areas."

The article concludes by stating that:

"These issues will persist throughout the next century and dominate major technical, scientific and political considerations into the indefinite future."

Although predicting the weather is a risky business at best, most people have had the experience of getting into a hot car with the windows rolled up. It is becoming increasingly clear that major climatic changes are already impacting many countries with severe droughts, hurricanes, forest fires and crop losses in the lower latitudes. According to insurance industry statistics, more weather related insurance claims were paid out in the first eight months of 1998 than during the entire decade of the 1980s. In addition, there were abnormal increases in the number of very large icebergs that have been breaking away from the polar icecaps. While it is not possible at present to prove conclusively that the current global climatic changes are caused by the greenhouse gases, it is hard to imagine that the three *billion* tons of carbon dioxide that are being added to the Earth's atmosphere each year will not have a profound impact on the Earth's climate.

Our human ancestors lived at a time when concern about the natural environment essentially did not exist as a priority. If anything, the natural world was something to clear and conquer. Indeed, the natural environment was typically viewed as an inexhaustible resource. However, the awesome power of exponential growth has turned a once-pristine planet into what has increasingly become a contaminated industrial wasteland. In many respects, the struggle is about over. The bulldozer culture is now in the mopping-up stages. The remaining ancient wilderness habitats are now few and far between and before long, they too will be gone. Despite this grim scenario, there is reason to believe that humanity is as close to utopia as it is to oblivion. While the global environmental problems are awesome, so is the collective potential of the global scientific and engineering community. It must, however, be assumed that due to the massive damage already inflicted on the Earth's biosphere, a series of ecological disasters are now inevitable. This is an important point because if one is aware that certain food-production systems are going to fail, action can be taken to minimize the losses. But the hour is late and there is no time to waste.

Trends

The current trends give reason for concern. Essentially, the problem can be reduced to more and more people competing for fewer and fewer resources. The problem is global, and grows more serious with each passing day. For many of the millions of starving people in the world, it is already too late. Temporary food can be provided for the short-term, but it will do little but prolong the inevitable. In many cases, the additional food means more babies will survive only to starve later. The question is: *How much time do those who live in the industrialized world have before they too will be threatened by an economic and/or environmental systems collapse?*

While no one can accurately predict exactly when such economic and ecological disruptions will occur, it is clear that the problems are growing more serious with the passage of time. The situation is similar to an earthquake. Before an earthquake can occur, tremendous pressure must first be built up over long periods of time in the geologic plates. The longer the pressure increases, the more devastating the earthquake will be. In the same way, enormous pressures have been building for many decades on the Earth's limited environmental resources; and sooner or later, their ability to sustain us will collapse. However, unlike an earthquake, the environmental and agricultural systems collapse that will be inevitable if nothing is done, can -- and should be -- directly affected by human action and planning. It is not necessary to perish like so many lemmings rushing into the sea. There are alternative paths available, but the necessary changes need to be made before the impending environmental dislocations foreclose our options to act rationally.

While short-term "band-aid" solutions that involve energy conservation, such as developing more efficient automobiles and appliances, will help to buy some time; ultimately such conservation measures will only prolong the inevitable. If new renewable sources of energy are not developed on a scale to displace the use of fossil and nuclear fuels, conserving what is left has been described as somewhat like taking a slow walk down a dead-end street. In addition, in the U.S., the progress in reduced energy consumption has to a large extent been the result of closing down the energy-intensive factories that were once the backbone of American industry. It is no secret that if the U.S. imports its steel, aluminum, copper, consumer electronics and automobiles, it will use less energy. The cost, however, has been to structurally displace millions of workers.

Time is of the essence. Because of the nature of exponential growth, if responsible action is not taken soon, the oblivion scenario may be closer than anyone suspects. Each generation throughout history has usually been faced with life-and-death problems and decisions, but due to the fact that many of the most serious energy and environmental problems are degrading exponentially, the magnitude of the problems now confronting the human community are unprecedented. Fortunately, the knowledge to deal with these problems is also increasing at a very rapid exponential rate. It is for this reason that humanity is now on the threshold of both utopia or oblivion, and the decisions that are made -- or not made -- in the next few years could determine which outcome will occur. One can only hope that there is still sufficient time to take the constructive actions that are necessary to protect the biological life-support systems of the Earth and the miracle of civilization that it has allowed to evolve.

Conclusions

Writing in *Scientific American*, William D. Ruckelshaus, the first administrator of the Environmental Protection Agency, stated that if humanity is to avoid the many environmental disasters that now appear to be inevitable, it will require *"a modification of society comparable in scale to only two other changes: the agricultural revolution of the late Neolithic Age and the Industrial Revolution of the past two centuries."* While the first two revolutions were *"gradual, spontaneous and largely unconscious,"* Ruckelshaus points out that the Environmental Revolution, *"will have to be a fully conscious operation, guided by the best foresight that science can provide . . . an undertaking that would be absolutely unique in human history."*

In the remaining chapters of this book, the specifics of how such a global energy and environmental revolution can be accomplished by making a transition to renewable energy resources will be described in some detail. If such a transition is successfully implemented, our reward will not be insignificant. For by allowing civilization to continue, we will ensure that the stunning developments in computer science, biotechnology and molecular medicine will continue. This will ultimately lead to a biological transition to renewable resources, as molecular biologists continue to unlock the secrets of life and the nature of disease by more fully understanding the specific molecular mechanisms involved. These molecular processes have been utilized by the vast and ancient microbial and protein-based "nanobial" organisms that are at the heart of metabolism, and of life itself. These biological "founding fathers," whose ancestors are more than three billion years old, will be discussed in more detail in Chapters 5 and 8, although their true significance to the life that exists on the Earth is only beginning to be fully understood.

In the next chapter, the critical nature of exponential growth and its relationship to the global energy and environmental problems will be reviewed. Few people have a even a basic understanding of exponential growth, and as a result, it is virtually impossible for one to understand the promise of the future, or the serious nature of the global environmental problems that now threaten that future. As such, there is no more important concept to comprehend than the nature of the "exponential age" in which we find ourselves.

Figure 2.1: *The Titanic*

CHAPTER 2

EXPONENTIAL ICEBERGS

The ocean liner *Titanic* was one of the major engineering milestones of its era. It was an elegant expression of what was technically possible, and it was thought to be unsinkable. Perhaps the most important lesson to be learned about the sinking of the *Titanic* is that it need not have happened. It has been reported that the captain of the *Titanic* knew full well there were icebergs in the area because his engineers on board had received numerous warnings from other ships that had reported seeing icebergs nearby. The other ships had either slowed or stopped, but not the *Titanic*, whose captain was apparently more concerned about setting an impressive transatlantic crossing record. This reckless overconfidence was the first major mistake.

The second major mistake was also the result of overconfidence. Because it was assumed that the *Titanic* was unsinkable, insufficient numbers of lifeboats had been placed on board. As a result, even after the *Titanic* had hit the iceberg and the passengers were well aware that the ship was going to sink in the icy waters, most of them could do nothing except wait for the inevitable. In the final hours that remained, many of the doomed passengers changed into their best formal wear and tried to enjoy their last minutes as much as they could under the circumstances. One might conclude from this experience that Mother Nature is very unforgiving of those who miscalculate. The more important lesson, however, is that the sinking of the *Titanic* was a catastrophe that could have been avoided if prudent actions had been taken in time.

It is the central thesis of this book that the integrated biological life-support systems of the Earth are very analogous to the *Titanic*. While most people assume that the Earth is unsinkable, there are a great many atmospheric and environmental scientists from around the world who have been issuing ominous warnings about multiple "icebergs" ahead. It is important to heed the warnings because the ever-accelerating environmental problems have the potential to jeopardize the global life support systems of the entire planet. The global contamination problem is somewhat like a nuclear war occurring in slow motion. And because the global environmental problems are compounding at an "exponential" rate, the time to take corrective action is increasingly limited. The passage of time is, therefore, a pressing concern. The question is: *How much time do we have?* To obtain an accurate "fix" on our position, it is necessary to understand the characteristics of exponential growth outlined in this chapter. Indeed, one cannot appreciate the genuine sense of urgency until one understands the simple but generally unknown exponential concept of "11:59."

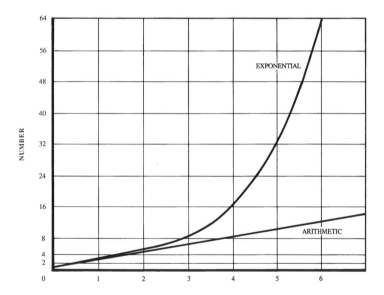

Figure 2.2: Arithmetic vs. Exponential Growth
Note that exponential growth is characterized by a curve that continues to increase until it goes vertically off of the page.

Albert A. Bartlett

One of the foremost experts on exponential growth is Albert Bartlett, professor of physics at the University of Colorado. Bartlett has published numerous papers and articles on exponential (also called geometric) growth and its relationship to the consumption of energy and environmental resources. Bartlett has emphasized that one of the greatest shortcomings of most elected officials is their inability to understand the exponential function and, in particular, how it relates to the global energy and environmental problems. Barlett describes exponential arithmetic as probably the most important mathematics students will ever come to understand. The importance of the exponential function is that it calculates steady growth. It is especially important to understand how a change in the sign of the exponent can make an enormous difference in the sum being calculated. What follows is a brief summary of Bartlett's key fundamentals of exponential growth that are described in his paper, "Forgotten Fundamentals of the Energy Crisis," published in the September 1978 issue of the *American Journal of Physics*.

When a quantity, such as the rate of consumption of a resource, is growing at a given percent per year, the growth is said to be "exponential." The important property of the growth is that the time required for the growing quantity to increase its size by a fixed fraction is constant. For example, a growth of five percent (a fixed fraction) per year (a constant time interval) is exponential. It follows that a constant time will be required for the growing quantity to double its size (increase by 100%). This time is called the doubling time, T2, and it is related to P, the percent growth per unit time by a very simple equation:

$$T_2 = \frac{70}{P}$$

A growth rate of five percent per year will result in the doubling of the size of the growing quantity in a time T2 = 70/5 = 14 years. But in two doubling times (28 years), the growing quantity will double twice (quadruple) in size. In three doubling times, its size will increase eightfold, and in four doubling times it will increase sixteen fold, etc.

Figure 2.3: Exponential Doubling Times
Each square represents a decade, and the area of each sqaure represents the quantity of petroleum consumed during that time. The area of
rectangle ABCD represents the known world petroleum resource.

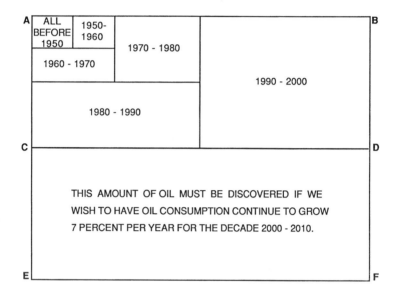

A simple but graphic example of exponential growth is to
consider what happens if one penny is saved on the first day of
the month, and each day thereafter, the amount is doubled. Few
people realize that at the end of a thirty-one day month, over ten
million, seven hundred and thirty thousand dollars would be
saved. Observe what happens to the numbers over time in Table
2.1. If a person's personal savings program were based on an
arithmetic increase, the increases would occur at a fixed amount,
rather than an amount that increases at an increasing rate. In
Table 2.1, for example, the amount saved each day is increased
by a constant arithmetic rate of $0.01. Note that at the end of the
31-day month, only $0.31 has been accumulated. This is in sharp
contrast to the $10,737,418.24 that would result if the amounts
were increased exponentially as they are in Table 2.2. An important characteristic of exponential growth and doubling times is
that the increase in any doubling is approximately equal to the
sum of all the preceding growth.

Table 2.1: Arithmetic Savings

Day	Amount($)	Day	Amount($)
1	.01	17	.16 + .01 = .17
2	.01 + .01 = .02	18	.17 + .01 = .18
3	.02 + .01 = .03	19	.18 + .01 = .19
4	.03 + .01 = .04	20	.19 + .01 = .20
5	.04 + .01 = .05	21	.20 + .01 = .21
6	.05 + .01 = .06	22	.21 + .01 = .22
7	.06 + .01 = .07	23	.22 + .01 = .23
8	.07 + .01 = .08	24	.23 + .01 = .24
9	.08 + .01 = .09	25	.24 + .01 = .25
10	.09 + .01 = .10	26	.25 + .01 = .26
11	.10 + .01 = .11	27	.26 + .01 = .27
12	.11 + .01 = .12	28	.27 + .01 = .28
13	.12 + .01 = .13	29	.28 + .01 = .29
14	.13 + .01 = .14	30	.29 + .01 = .30
15	.14 + .01 = .15	31	.30 + .01 = .31
16	.15 + .01 = .16		

Note on day 10 of Table 2.2, for example, that the $5.12 saved is twice the amount of the total savings ($2.56) of the previous 9 days. Another important aspect of exponential growth is that even modest rates of growth can still eventually result in enormous consequences. For example, world production of crude oil increased only at a rate of about 7 percent per year from 1870 to 1970, but a 7 percent rate of growth means the doubling time is only 10 years. The danger of exponential growth is that when it first starts to grow, the level of growth seems insignificant, which leads to complacency. But after a few doubling times, even small amounts can increase into staggering quantities. Even relatively small annual increases of one or two percent can be significant. Since 1900, for example, the human population has been increasing at approximately two percent per year. Yet, even at that modest level of growth, the human population has increased from approximately 1.6 billion to over 6 billion in the 100 years between 1900 and the year 2000.

Table 2.2: Exponential Savings

Day	Amount($)	Day	Amount($)
1	0.01	17	655.36
2	0.02	18	1,310.72
3	0.04	19	2,621.44
4	0.08	20	5,242.88
5	0.16	21	10,485.76
6	0.32	22	20,971.52
7	0.64	23	41,943.04
8	1.28	24	83,886.08
9	2.56	25	167,772.16
10	5.12	26	335,544.32
11	10.24	27	671,088.64
12	20.48	28	1,342,177.28
13	40.96	29	2,684,354.56
14	81.92	30	5,368,709.12
15	163.84	31	10,737,418.24
16	327.68		

The Concept of 11:59

In order to comprehend the sense of urgency with regard to the problems that now confront the human community, one needs to understand the exponential concept of "11:59." As Bartlett points out, there are many different examples that can be used to explain the concept of 11:59. Some of the more common examples include lily pads growing in a pond, the number of people increasing on the Earth, or bacteria growing in a bottle. Bacteria, for example, grow by division so that one bacterium becomes two and the two divide to give four, etc. Assuming the division time is one minute, the bacteria will be growing exponentially with a "doubling time" of one minute. Given this scenario, if one bacterium is put in an empty bottle at 11:00 in the morning, and it is observed that the bottle is full of bacteria at 12:00 noon, consider the following question:

When was the bottle half-full?
Answer: 11:59

If you were one of the bacteria in the bottle, at what point would you first realize that you were running out of space and therefore "resources?" Consider that at 11:55, the bottle is only 3 percent filled, which leaves 97 percent open space -- just waiting for "development." As Bartlett likes to point out, suppose that at 11:58, some farsighted bacteria realize that they are rapidly running out of resources, and with a great expenditure of effort they launch a search for new bottles. Let us assume that as a result of their efforts, the bacteria are fortunate enough to discover three new empty bottles. Great sighs of relief would be expected to come from all the worried bacteria. But even with the total space resource quadrupled, the exponential growth of the bacteria could only be maintained for another two minutes! Table 2.3 summarizes what happens to the numbers over time:

Table 2.3
Bacterial Exponential Growth

11:55 a.m.	Bottle #1 is 3% full.
11:58 a.m.	Bottle #1 is 25% full.
11:59 a.m.	Bottle #1 is 50% full.
12:00 noon	Bottle #1 is 100% full.
12:01 p.m.	Bottles #1 and #2 are full.
12:02 p.m.	All 4 bottles are full.

11:00   11:55   11:58   11:59   12:00

Figure 2.4: 11:59

Figure 2.5: Human Population Growth
It should be obvious to anyone looking at this graph why the existing human population growth is unsustainable. After a certain point, note how the graph goes vertically off the page.

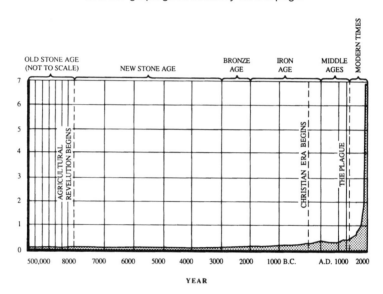

The Age of Exponentials

We are living at a time when the exponential increases in a large number of key environmental areas are (to borrow a term from the nuclear physicists) going critical. That is to say, in many areas, after a certain point in time, it will no longer be possible to alter the outcome. The primary concern is what happens when the existing fossil fuel resource capabilities of the Earth are consumed by the exponential human population growth. The answer can be referred to as a "systems collapse," whereby population levels are catastrophically reduced. A more common term is "die-off," but whatever one calls it; it is serious business. As the passengers on the *Titanic* found out, Mother Nature, which is essentially the laws of physics and chemistry, has consistently been shown to be extremely unforgiving to those who violate the laws by exceeding the physical limits of the ecosystem in which they live. Thus only rational planning and actions will allow the passengers of *Spaceship Earth* to avoid the fate of the *Titanic*.

Progress is generally viewed as a good thing, but given the nature of exponential growth, it would appear that what the human community is progressing toward with ever-increasing speed are numerous "exponential icebergs." Key problem areas (icebergs) that have been deteriorating exponentially for many years, if not decades, include the following:

Human population growth
Consumption of nonrenewable fossil fuels
National debt
Acid deposition
Carbon dioxide accumulations in the atmosphere
Stratospheric ozone depletion
Production of toxic chemicals
Destruction of forests (deforestation)
Advancing of the deserts (desertification)
AIDS virus afflictions
Poverty
Destruction of marine and wildlife habitats

The long-term impact of the exponential consumption of conventional fossil fuel resources will be discussed in more detail in the next chapter. However, the key point to understand is that oil accounts for the vast majority of energy consumption in the U.S. (as well as most of the rest of the world) and the estimates from the American Petroleum Institute and the U.S. Geological Survey indicate that with without oil imports, at the current rates of consumption, the U.S. will exhaust its current oil reserves (which represent only about four percent of the remaining global reserves) in about 10 years. Known world oil reserves will be exhausted between 2050 and 2080. While additional oil will surely be discovered, at present, the world is only producing one barrel for every *four* it consumes. This being the case, it will be increasingly expensive to extract the remaining oil reserves. Even if the supply of oil and other fossil fuels were unlimited, their continued use as a combustion fuel could have a catastrophic impact on the Earth's ecological life support systems. As a result, the U.S. and other industrialized countries should be moving with a sense of urgency to make a transition to renewable resources. In spite of these global issues, as of yet, no coherent strategy in the U.S. has yet evolved to deal with them.

Figure 2.6: The Iceberg

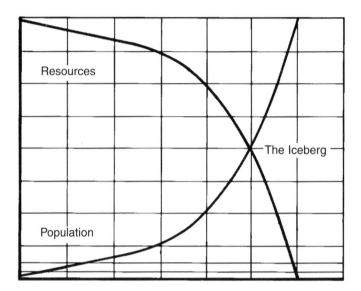

Positive Exponentials

It is important to realize that exponential growth does not have to be negative. It all depends on what is growing exponentially. An exponentially increasing savings account is obviously a highly desirable asset to own. As financial planners are fond of telling their clients, even a one or two percent increase in the annual rate of interest can make an enormous difference in one's savings over a period of ten or twenty years. Although the exponential growth in knowledge and information is much more difficult to quantify, there is no question that it is increasing at a staggering rate. This is particularly the case in the technical areas of engineering, computer science, molecular biology and medicine. It is only because of these explosive developments in information and technology that it seems reasonable to expect that molecular medicine will soon usher in an age where molecular disorders and the diseases they cause will essentially be eliminated. With such fundamental changes in human life span potential, a future of unlimited possibilities awaits those who will be able to take advantage of such biotechnologies.

Such options are based, however, on an assumption that the human community will be able to survive the many serious global environmental problems that continue to degrade exponentially. Never before has humanity been at the crossroads of such awesome opportunities -- or problems -- as the exponential forces of life and death are simultaneously evolving and racing toward some ominous conclusion. What needs to be understood by every thinking person is that the decisions that are made now could well make the critical difference in the ultimate outcome. This places an extraordinary responsibility on the print and electronic news media to communicate the fundamentals of both the problems and solutions to the general public -- while there is still time to take corrective actions.

The Information Explosion

Of all of the phenomena that are increasing exponentially, none is more significant -- or more promising -- than the increases in knowledge and information. Carl Sagan, professor of astronomy and space sciences at Cornell University, summarized the unprecedented information explosion in his book *Cosmos*:

> *"The great libraries of the world contain millions of volumes. If I finish a book a week, I will read only a few thousand books in my lifetime, about a tenth of a percent of the contents of the greatest libraries of our time. The trick is to know which books to read."*

The rates at which knowledge, and its direct spin-off, technology, have been increasing is probably incalculable. Even to try and stay current in one specific area of medicine or engineering is exceedingly difficult, if not outright impossible. As a result, the most significant question to be asked in this "Age of Exponentials" with its seemingly endless sea of information is: *What is worth knowing?* Ultimately, it is a question of priorities, and priorities to a large extent depend on one's awareness of what is happening in one's immediate environment. If one's house is burning down, priorities are easy to establish because of the immediate and obvious threat.

However, problems dealing with acid deposition, radioactive waste, or ground water contamination are complex and the threat is not immediately observable. Moreover, effective solutions, if they exist at all, are usually very expensive. This adds an additional dimension to the problem. Specifically, *who is going to pay to resolve the problem?* Such complex issues are confusing even for highly trained specialists -- much less members of the general public. This no doubt explains why such large numbers of people feel helpless to do anything. Because they cannot even begin to cope with the tide of events sweeping over them, the understandable response is to become apathetic and "tune out." Given this perspective, it is easy to understand why depression, mental illness and drug addiction problems are so widespread, as increasing numbers of people find they are unable to cope with a high-pressure, high technology society that continues to change at a very rapid exponential rate.

Future Shock

The concept of "future shock" was first described in 1965 by Alvin Toffler, who later published a book, *Future Shock*, which defined the condition as the shattering stress and disorientation that can be induced in individuals by subjecting them to too much change in too short a time. In some societies, centuries can pass with virtually no significant changes taking place. But in contemporary urban-industrial societies, the exponential acceleration of change has, in and of itself, become "an elemental force," and increasing numbers of people are simply unable to emotionally adapt to it. As Toffler wrote:

"Whether we examine distances traveled, altitudes reached, minerals mined, or explosive power harnessed, the same accelerative trend is obvious. The pattern, here and in a thousand other statistical series, is absolutely clear and unmistakable. Millennia or centuries go by, and then in our own times, a sudden bursting of the limits, a fantastic spurt upward."

Toffler then asserts that this acceleration of change has now reached a point where it can no longer be regarded as "normal." Yet, the rate of change continues to accelerate.

What becomes increasingly obvious is that technology seems to have a life of its own. It is important to remember, however, that the information explosion and the rapid technological change it has stimulated is as much a part of the solution as it is the problem. Indeed, there are many people who seem to be thriving in this era of rapid change. Such individuals are not concerned about the future arriving too soon; they wish it would hurry up. Of course, such people usually have the general assumption that things are going to get better in the future, as indeed they may. But while it is important to keep a positive outlook, it is equally important not to forget about the problems that are compounding exponentially because if we do not deal with them, they are going to deal with us.

Strategies for Survival

1. It is important to be part of the solution and not part of the problem. Not being aware of the problems will not make them go away. Rather, the exponential problems must be dealt with while there is still time to take corrective action.

2. It is important to understand that the situation is not hopeless. Humanity is as close to utopia as it is to oblivion. This is no time to give up. While it is true that the problems are unprecedented, so are the technical resources that are available to solve these problems - if we choose to use them for such purposes.

The solution ultimately involves the industrialized nations of the world making a transition from nonrenewable to renewable energy and biological resources. This "transition of substance" is a political problem, and political problems require political solutions. What is required is a reordering of the national and state priorities. Before that can happen, however, the human community must be made aware of the magnitude of the problems, and the realistic options that are available. The problem is that most of the options involve technical issues and questions that place an in-creasing emphasis on the education of not only the young, but everyone.

Education: Problems and Solutions

Education is the foundation of political action. But given the nature of the information explosion, the role of education takes on a new meaning. It is no longer just a question of absorbing information, because one can only assimilate a small fraction of that which is available. As a result, the most important educational decisions an individual can now make are those which revolve around the question of *what is worth knowing?* There have been numerous news accounts of reports by various experts in education who have been critical of the U.S. educational system, particularly in the primary and secondary levels. One such report was prepared by Glenn T. Seaborg, a Nobel Laureate who has been a technical advisor to various presidents. Seaborg is a former member of the Manhattan Project during World War II, and he served as National Chairman for the development of high school chemistry courses. He is currently a researcher at the University of California, and in 1983, he was appointed as a member of a special national commission that was to investigate the state of American education. After many months of research, the commission concluded the following:

> *"If an unfriendly power had attempted to impose upon the United States the mediocre educational performance that exists today, we might have viewed it as an act of war."*

In the 1988 Public Broadcast System (PBS) television program *Innovation,* narrator Jim Hartz interviewed Seaborg who indicated that the situation is not getting any better, primarily because students are not being "turned on" by teachers that are often not interested in, or well-trained in, science and mathematics. When asked, "If it takes inspirational teachers to bring students along, and we are not producing very many, what's the answer?" Seaborg replied, "That's a very difficult problem." He went on to explain that it is a matter of restoring elementary and high school teachers to the status that they used to enjoy -- and that they still do enjoy in countries such as Japan and Germany.

When the U.S. educational system is compared to the one in Japan, there are many sharp contrasts. Per-capita student expenditures in Japan are reported to be less than those of U.S.

students, yet according to a 1988 CBS 60 Minutes broadcast titled "Head of the Class," the average Japanese student attends school 240 days per year, compared to 180 days for the average American student. In addition, the average American student's school day starts at 8:00 a.m. and ends at about 3:00 p.m., whereas the average Japanese student begins at about 7:00 a.m. and continues until midnight. While there are many Americans who would understandably be opposed to such a demanding system, there is the hard reality that it is not just the Japanese educational system that the U.S. is lagging behind. America's educational system compares unfavorably to those in virtually all of the industrialized countries and even many Third World countries.

Perhaps the most obvious solution would be to have the U.S. educational system adopt comparable academic standards to those in Japan and Europe for each grade level. If the students do not pass a course with at least an 80 or 90 percent proficiency, they are simply not allowed to progress to the next level. The concept of "social passing" should be eliminated. After all, it is what you know, and not how old you are that is important. Part of the problem, however, is a pervasive attitude of many American students (and adults) that can be reduced to the reality that they "live to play." This attitude begins early in life, as children in contemporary American culture are taught to play instead of using their time in a more constructive manner during their most formative and impressionable young years.

Starting Young

It is important to realize that mastering a scientific discipline initially involves learning to speak its language, and it has been observed that young children are able to absorb information or acquire languages much easier than their adult counterparts. If this is the case, it is especially important not to let young children idle away a critically important part of their formative years. A generation ago, the typical American child who grew up on a farm was usually given important responsibilities at an early age. This is still true in much of the Third World. In sharp contrast, however, U.S. children are rarely exposed to serious responsibilities until after high school, or in many cases, college.

Consider how much time many high school students invest in playing football or baseball. If the same effort were invested in mathematics, chemistry, biology or physics, American students would be well on the way to being competitive with their Japanese or European counterparts. This is not to suggest that students should not participate in athletic events, but some sort of perspective needs to be maintained. It is advisable to reexamine the notion of encouraging students to spend thousands of hours developing short-term athletic skills, rather than long-term intellectual skills.

It is also important for each individual to observe how they are using their "free" time. This free time is an incredibly valuable resource that many people just idle away. If it is used wisely, it compounds over the years like interest in the bank. This, however, is a lesson that is rarely taught in most classrooms. What is especially onerous is the exponentially increasing levels of violence that is finding its ways in to motion pictures -- and the fact that millions of highly- impressionable young people spend long hours watching and re-watching such "entertainment." It is not a question of the violence that one is exposed to in movies such as *Saving Private Ryan*, but rather the new generation of *Chainsaw Massacre*-type moves where the violence has reached unprecedented levels of depravity. It is these "hyperviolent" films that entrance millions of young people every weekend, and one can only wonder how such movies will impact these young viewers during their formative years. If a comparable expenditure of effort were devoted to intellectual skills in chemistry or mathematics, one would have knowledge that would be useful throughout one's life. *Ah, but chemistry and thermodynamics are boring, right?* Wrong! If there are people who think such fascinating subjects are boring, they know nothing about such subjects.

Teaching Teachers

While the attitude of students and parents is certainly an important factor in one's educational development, there is no question that the serious problem raised by Seaborg is of paramount concern; namely, that there are few primary or secondary teachers that are equipped to teach technical subjects at all, much less in a creative and inspiring way.

There is, however, a way out of this maze. While there may not be many teachers who are particularly gifted in teaching mathematics or science, there is no question that there are more than a few. This being the case, and given the miracle of video-tape, it is certainly possible to have production quality films and/or videotapes produced and shipped to every school in the country, on every major area of scientific investigation.

Such video programs, which could and should be funded by the federal government, would allow the poorest inner city or rural school children to have access to the very best instructors that exist in the country -- or the world for that matter. In addition, the programs could be made in such a way that the teacher is not just standing in front of a classroom lecturing. Rather, *"Star Wars"* quality modeling and computer animation could be incorporated to make learning and education truly entertaining, which it should be anyway. There is a special bonus in this approach. As the teachers are showing the high-quality videotapes to their students, the teachers themselves will soon be able to master the material presented, and thereby have their own education significantly upgraded while "on the job." As any teacher knows, the best way to master any subject is to teach it.

Another important bonus to this approach is that if a student had trouble understanding all of the material during class (or perhaps the student was out ill), he or she can take the video home, and watch it in the evening as many times as it takes. This not only provides for individualized learning, but perhaps the family can be present so learning can become an enjoyable family affair. This will also provide parents with an opportunity to gain a better understanding of what their children are being taught in school, and allow the parents to upgrade their education as well.

There have already been many high-quality documentaries developed by the Public Broadcasting System network, including the *Nova* and *Frontline* series. What is tragic is that so few people watch these consistently outstanding productions. This reinforces the need to consider carefully how young people, as well as adults, use the limited amount of leisure time that they have available. It appears that a great many people spend their free time reading low-grade fiction (assuming they read at all), or watching televised entertainment programs, such as soap operas or television sitcoms, in contrast to the documentaries that could help them make more informed judgments.

Multiple Exponential Problems

Given the serious nature of the complex and interrelated problems that are continuing to compound exponentially, the net result of this unawareness could well be catastrophic. The exponential age in which we now live is rapidly approaching both the oblivion and utopia scenario. It is unfortunate that the nature of exponential growth is not understood by most people. That will not, however, alter its inevitable result. Moreover, the dilemma is further compounded by the fact that the U.S., and the rest of the world, is not just threatened by one exponential problem, but many. It is critical that the majority of people understand both the nature of the problems -- as well as the solutions -- that are presently available.

In this regard, the process of communication and education is absolutely essential if there is to be any chance of implementing the necessary changes in time. We can only hope that it is not already too late to avoid the more serious environmental problems that not only threaten human civilization, but the Earth's biological life-support systems that have been evolving for literally billions of years. Those who attempt to detach themselves from the actuality that threatens everyone are a major part of the problem. This detachment is a delusion, because the Earth's biological life support systems are swiftly approaching the crossing of the exponential curves whereby the ascending line of consumption will intersect the descending line of available resources. If such a progression is allowed to continue, in the end, there will be no place to hide. The urgency of the situation is due to the massive inertial effects that are a direct result of exponential growth.

Even though the captain and at least some members of the crew of the *Titanic* had found out about the iceberg well before the point of impact, they did not have sufficient time to change course because of the enormous momentum the ship had built up. In a similar context, the global energy, economic, and ecological problems that have been accumulating gradually over many decades are not going to be responsive to sudden changes. Moreover, there is a point-of-no-return beyond which changing course is simply not possible. If this generation is to act rationally, it must not be complacent about the exponential events that are unfolding.

Conclusions

Perhaps the most important lesson of exponential growth is that it is later than we think. In the next two chapters, the impact of exponential growth on conventional fossil fuel and nuclear resources will be discussed in more detail. It is important to keep in mind the fact that making fundamental changes in something as basic as the global energy infrastructure is going to take decades. Moreover, as the fossil fuel and uranium reserves are consumed, it is only reasonable to expect their prices to increase as the global environmental problems continue to escalate.

Such is the nature of the problems that now confront the global human community. We are not talking about minor considerations. We are talking about a potential global systems collapse on a scale that is difficult to comprehend. But comprehend we must if we and our children are to be survivors. The most important consideration is that there are viable alternatives to the fossil fuel and nuclear energy systems, and that the primary obstacles to mass-producing such renewable energy systems are not technical, or even economic. In the final analysis, *they are educational and political.*

Figure 3.1: Offshore Drilling
Finding the last of the last of the oil.

CHAPTER 3

FOSSIL FUELS

"Today, the world finds one new barrel of oil for every four it consumes. World oil discoveries peaked in the 1960s; in the United States, they peaked in the 1970s. More than 90 percent of today's oil comes from fields discovered more than 20 years ago - and most of the fields uncovered in the past decade have been extremely small."

Popular Science
May 2000

Interrelationships

Many of the global environmental and economic problems are directly related to what types of energy are used and how that energy is generated, transported and stored. Some of the obvious environmental concerns discussed in Chapter 1 include global climate change, as well as ground, air and water pollution. Energy choices impact acid deposition, carbon dioxide accumulations, oil spills, gasoline seepage from storage tanks, strip-mining, preservation of the remaining wilderness areas and the production of radioactive wastes. Energy choices also impact significant economic factors, including inflation, interest rates, unemployment, and both domestic and foreign trade deficits. These interrelationships should not be surprising given the fact that environmental, economic, political, military, social and industrial systems are all dependent upon reliable energy resources and conversion systems. As such, they are secondary systems that are all predicated on a primary energy system.

It follows then, that if the basic energy system is in trouble, all of the secondary systems will be also. To concentrate only on the secondary systems is like treating the symptoms of a disease rather than the cause. Edward H. Thorndike, a professor of physics at the University of Rochester, has made the observation that although increasingly scientists and engineers are being called upon to give advice on energy questions, few of them have adequate knowledge that extends beyond their own areas of specialization. This is compounded by the fact that environmental and energy problems have extensive interconnections, which means a broad knowledge of the "big picture" is at least as important as a detailed knowledge of any one aspect of it.

This is no small consideration. There is a general principal in education whereby as one advances, one's area of specialization invariably narrows. This results in a great many highly educated specialists who know a great deal about their particular area of interest, but little or nothing about the infinite number of other subjects that they simply don't have time to investigate. The comprehensive analysis and communication of data can be a critically important function, yet few professionals are able to follow the events occurring outside their immediate work. This communications problem is a principal obstacle that has thus far prevented the organization and implementation of a global reindustrialization effort around renewable energy resources.

The individuals who think the energy crisis is over do not realize that everything they now buy has been impacted by the dramatic increases in oil prices that occurred in 1973 with the Arab oil embargo. Prior to the embargo, oil was selling for about three dollars per barrel. After the embargo, the average price of oil quadrupled to about twelve dollars per barrel. Although oil prices have fluctuated dramatically since the 1970s, ranging from fourteen to in excess of thirty dollars a barrel, oil prices have never returned to their pre-embargo levels, in spite of a recent worldwide oil surplus that occurred in the 1980s and 1990s.

If a person buys a new automobile, the average price is now about $20,000, whereas the same basic vehicle could have been purchased in 1970 for about $3,000, which is an increase of over 650 percent. Another obvious example is the cost of housing. Homes that used to sell for $25,000 before the oil price hikes in the 1970s now cost around $150,000. Numerous other examples listed in Figure 3.2 illustrate how a wide-range of product prices increased with the price of oil. The examples in Figure 3.2 illustrate how the purchasing power of the U.S. dollar has fallen dramatically in the thirty-year period from 1970 to 2000. While the average cost of living increased by roughly 600 percent, in many cases personal income actually declined.

Figure 3.2: Inflation
Costs of Goods from 1970 to 2000

Examples	Average Costs In:		Percent Increase
	1970	2000	
Monthly Housing	$150.00	$1000.00	660%
Monthly Car	85.00	425.00	500%
Loaf of Bread	.25	1.50	600%
Coffee (1 lb)	.50	3.00	600%
Candy Bar	.10	.50	500%
Man's Dress Shirt	5.00	28.00	560%
Gasoline (per gal.)	.25	1.50	600%
Monthly Electric Bill	100.00	400.00	400%
College Tuition	300.00	1,800.00	600%
Postage	.05	.33	660%

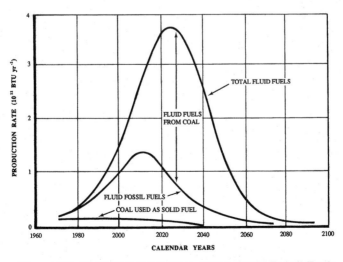

Figure 3.3: Projected Production of the World Fossil Fuels
Prepared by Department of Physics,
Texas Tech University, Lubbock, Texas.

Every product produced has a fundamental energy cost factor. When the cost of energy increases, inflation increases, and since interest rates must always be several points higher than the rate of inflation, interest rates go up as well. This explains why both inflation and interest rates came down with the price of oil in the 1980s, in spite of record federal deficits. But as Figure 3.3 illustrates, the current oil glut is temporary. The science editors of *Popular Science* summarized the situation in the May 2000 issue as follows in their article: *Are We Really Running Out of Oil?:*

> *"Suffering from sticker shock at the pump? Don't expect a reprieve anytime soon. Though the most recent increase in U.S. gasoline prices was the result of a decrease of oil production among the 11 OPEC (Organization of Petroleum Exporting Countries) members, the world will be facing a much bigger issue over the next 10 years, one that will keep prices inflated: we're running out of oil.*
>
> *Really.*
>
> *Though the general public -- having been warned about the depletion of our oil reserves for decades -- is taking an "I'll believe it when I see it attitude," the facts are a little unnerving. Today, the world finds one new barrel of oil for every four it consumes. World oil discoveries peaked in the 1960s; in the United States, they peaked in the 1970s. More than 90 percent of today's oil comes from fields discovered more than 20 years ago -- and most of the fields uncovered in the past decade have been extremely small. The consensus in the industry is that, based on current reserves, anticipated discoveries, and today's rate of demand, the world will run out of its most precious resource in the year 2050. (OPEC, it should be noted, believes supplies will last until 2080.)"*

Although new oil reserves will certainly be found (mostly abroad), it is highly unlikely that new discoveries will keep pace with the increasing global demand. According to Joseph Riva, a petroleum geologist who studies the industry, *"We have a reasonable idea of where to find oil, and there just aren't that many areas left to explore."*

Energy Fundamentals

Because oil and other fossil fuels are nonrenewable, it is reasonable to assume that while their prices will vary with market forces, and their overall costs will continue to increase over time as their reserves are depleted. This, in turn, means that the longer a transition to renewable resources is delayed, the more expensive the transition will be. According to data provided by the U.S. Energy Information Administration in Washington, D.C., the principal sources of the world's energy in 1997 were liquid fuels (crude oil and natural gas liquids), natural gas, coal, and electricity from hydropower and nuclear power. Measured in British thermal units (Btu), the total world production of energy exceeded 381 quadrillion (i.e., quads) Btus. (Note: a quadrillion is a 1 followed by 15 zeros). Crude oil accounted for 142 quads, which is about 37 percent of the total. Coal production of 92 quads accounted for 24 percent of the total. Dry natural gas accounted for about 84 quads, or 22 percent of the total. Hydroelectric power accounted for 26 quads, which accounts for about 7 percent of the total, and nuclear power contributed 24 quads, or about 6 percent.

The leading energy producer in 1997 was the U.S., which generated about 73 quads and consumed 94 quads, requiring imports of 26 quads. Russia was the second largest energy producer, which generated about 40 quads of energy. China was the third largest energy producer at 36 quads, followed by Saudi Arabia at 21 quads. Although energy production of the Middle East countries is only about 13.6 percent of the world total, according to *Oil and Gas Journal*, the Middle East has roughly two thirds of the known global petroleum reserves, which are estimated to be 1,000 billion barrels.

All current and potential sources of energy can be classified as either renewable or nonrenewable. Nonrenewable sources include the fossil fuels, such as oil, coal and natural gas, while renewable sources generally refer directly or indirectly to solar energy. The one notable exception is geothermal energy, which is generated from the Earth's molten interior. Indirect solar resources include hydropower, wind, wave, ocean thermal energy conversion (OTEC) and biomass-fueled systems. Direct solar energy systems include photovoltaic cells, line or point-focus concentrator systems, and flat-plate collector systems that are used for water or space heating applications.

While the fossil fuels were also initially generated by solar energy processes, in that they are the products of green plants, the process of biological decay that formed the fossil fuels took millions of years. From a practical standpoint, therefore, fossil fuels are generally categorized as a nonrenewable resource. Although electricity accounts for about 30 percent of the U.S. energy consumption, it only provides about 10 percent of the net energy use. This is because roughly two-thirds of the energy at the power plants is lost as waste heat (over 22 quads annually), which only contributes to the global warming problem. Fossil fuels such as oil, coal or natural gas account for about 90 percent of current U.S. energy consumption. Even the bulk of electricity (about 70 percent) is generated from the burning of fossil fuels.

Coal

Coal is the most abundant, and from an environmental perspective, the dirtiest of the fossil fuels. In 1997, carbon dioxide emissions from coal plants reached half a billion metric tons of carbon, which is 32 percent of all the carbon dioxide emitted in the U.S. Coal also accounted for a third of all energy produced in the U.S. in 1997. Coal is the least expensive of the fossil fuels. According to the U.S. Energy Information Administration, in 1998, production prices for coal were 83 cents per million Btu (mmBtu) compared to $1.77 for natural gas and $1.88 for oil.

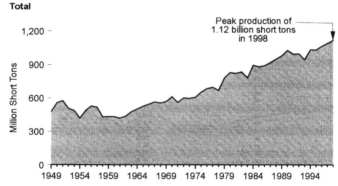

Figure 3.4: U.S. Coal Production
Comparing the years 1949 to 1998
Source: U.S. Energy Information Administration

In the U.S., coal provides about 57 percent of the energy consumed by electric utilities, natural gas provides about 10 percent and petroleum provides about 4 percent. Nuclear fission power plants generate about 20 percent of the electricity generated and hydroelectric dams provide about 10 percent. In the U.S., coal is the most abundant fossil fuel, and its reserves have been estimated by many "experts" to be sufficient to provide for the U.S. energy needs for roughly 500 years. However, this supposed long-term availability is based on the assumption that coal is used almost solely for electricity generation (as it is today). This, of course, implies that there will be some other vast source of energy that will be used to power automobiles, aircraft, and industries that use combustion fuels, and not electricity. But even if it were technically possible to mine and utilize enough coal to replace oil and natural gas, assuming an annual growth rate of 7 percent, the 500-year supply of coal would then be reduced to about 60 to 70 years,

In addition, according to Donald F. Othmer, an engineering professor at the Polytechnic Institute of New York (Brooklyn), a reindustrialization effort to convert coal, shale oil and tar sands into useable liquid fuels is a massive industrial undertaking. Othmer estimated that the engineering effort alone would require virtually all of the 25 or 30 thousand competent engineers in the U.S. working for 40 years and the overall cost would easily exceed a trillion dollars. In addition, the environmental impact of a major transition to coal (in terms of strip mining and carbon dioxide emissions) would be devastating.

Crude Oil Reserves

World crude oil reserves were estimated to be about 1,000 billion barrels as of 1997, and roughly 677 billion barrels of these reserves were found in the Middle East. The top five countries with the largest reserve totals (in billions of barrels), were:

1.	Saudi Arabia	261
2.	Iraq	112
3.	United Arab Emirates	97
4.	Kuwait	96
5.	Iran	93

These five Middle East countries account for roughly 70 percent of the world's crude oil reserves. The U.S. now ranks eighth in world crude oil reserves, accounting for only about two percent of the world total with about 22.5 billion barrels. Since U.S. consumption is about 2.2 billion barrels of crude oil annually, simple division indicates that if the U.S. did not import foreign oil, its own reserves of crude oil would be exhausted in about 10 years.

$$\frac{22.5 \text{ billion barrels}}{2.2 \text{ billion barrels}} = 10 \text{ years}$$

A great many numbers are used in discussions of energy reserves, and there is always the question of whose numbers are correct. In the past, it was a common practice for energy and natural resource analysts to tabulate all of the known oil reserves and divide the sum by the current rate of consumption to determine how long the resource would last. Using this method has resulted in many inaccurate forecasts. Predictions were made in the 1920s that the world's known oil reserves would be exhausted in 20 years. The world did not run out of oil in 1940 because new discoveries of oil increased the known reserve base. It is repeatedly assumed, based on experiences such as this, that as long as the free market is allowed to operate, new discoveries of oil will be able to continually offset the increasing levels of consumption.

One of the obvious problems with this estimation method is that the reserves used in the analysis were only those that were actually discovered. In contrast, when M. King Hubbert, a recognized world authority on the estimation of energy resources undertook his analysis of the world's fossil fuel reserves, he carefully estimated the ultimate total production of the fossil fuels (i.e., the total size of the resource that is recoverable). This "ultimate total production" of the resource will not essentially change, regardless of when or how many of the reserves are actually discovered. In addition, the ultimate total production will not be affected by free market forces or government action or inaction. It is for these reasons that Albert Bartlett and his colleagues at the Department of Physics at the University of Colorado have used Hubbert's data in calculating how long the U.S. and global fossil fuel reserves might be expected to last.

Exponential Expiration Time

Bartlett has been carefully analyzing energy and natural resource data for many years; and utilizing information from Hubbert and others, he has been able to calculate how long the U.S. and world fossil fuel reserves will be expected to last, given various rates of growth in consumption of the resources. One of the most significant aspects of Bartlett's calculations is what he refers to as the EET of a given resource. Bartlett emphasizes that growth is used as the primary indicator of economic progress. The growth of production, consumption and the Gross National Product (GNP) is the central theme of the U.S. economy, and it is regarded as disastrous when actual rates fall below projected levels. Because growth has become synonymous with success, it is important to calculate the life expectancy of energy resources given various assumptions about growth. Of special significance is the period of time that is necessary to consume a given resource, which Bartlett refers to as the "exponential expiration time" (EET) of the resource.

The EET is calculated by knowing the size of the resource, the rate of its use, and the fractional growth-per-unit time of the rate of consumption. While most resource scholars may be familiar with this equation, Bartlett has observed that there is little evidence that it is known or understood by the political, industrial, business, or labor leaders who continue to emphasize how essential it is for our society to have uninterrupted growth. Bartlett points out how important the relationship is between the lifetime of a resource and the rate of consumption. For example, a resource that will last for 1,000 years at current rates of consumption (i.e., zero growth) will be exhausted in 115 years if the annual growth increases by only 3 percent. According to data compiled by Hubbert, world oil production from 1890 to 1970 has grown at a rate of about 7 percent, which results in a doubling time of 9.8 years.

It is important to note that oil and other fossil fuels reserves will never be totally exhausted. It is just a question of how soon they will become too costly to extract and use what is left. According to an estimate that was issued by researchers at the U.S. Geological Survey (USGS), world oil reserves totaled only 723 billion barrels (i.e., about a 36-year supply). The report estimated that undiscovered resources would account for about 550 billion barrels, for a total of 1,273 billion barrels.

Other investigators have reached similar conclusions with respect to fossil fuel reserves. Two studies in particular are worthy of mention. The first was a two-year international effort under the direction of Carroll Wilson of the Massachusetts Institute of Technology (MIT). Some 70 energy analysts were recruited from government and the universities in fifteen major countries to focus on calculating energy supply and demand. Their conclusions of the studies were as follows:

1. There is only a finite amount of oil and there are limits to the rate at which it can be recovered. Sometime before the year 2000, the decreasing supply of oil will fail to meet the increasing demand.

2. Because large investments and long lead times are required for developing new energy resources, it is important that the effort begin immediately.

The second two-year-long study was undertaken at the request of the USGS and the Department of Energy by the RAND Corporation, a highly respected research group based in Santa Monica, California. The conclusions of the 700-page report were that the prospects of finding more oil and gas in the U.S. are severely limited. Moreover, the reason is geology, and not economics. The U.S. is simply running out of unexplored places where there is any possibility of finding significant amounts of oil.

However, even a doubling of the resource only results in a small increase in its life expectancy given the nature of exponential growth. This hard reality is not overlooked by senior executives of the major oil companies. In an essay published in *Newsweek* magazine, Allen E. Murray, then president of the Mobil Oil Corporation, repeatedly warned against complacency. He indicated that new oil shortages were inevitable, the only question was when. Other experts outside of the oil industry have made similar warnings about declining U.S. oil reserves. Calculations of world oil reserves are not encouraging, particularly in the face of growing levels of consumption by Third World countries which are only beginning to industrialize their societies. In sharp contrast to these perspectives are the beliefs of many people who assume that there are enough fossil fuel reserves to last for thousands of years. As Bartlett indicates, however, it is possible to calculate an absolute upper limit to the amount of oil the Earth could contain. He simply asserts that the volume of oil in the Earth cannot be larger than the volume of the Earth itself. The volume of the Earth is roughly equivalent to about (6.81 x 10^{21}) barrels. As a result, if the rate of oil consumption in 1980 were to continue with an annual increase of about 7 percent, an entire Earth full of oil would be consumed in only 342 years.

Petroleum Reserve Estimates

Geophysicists were predicting as far back as World War II that by the year 2000, U.S. oil reserves would be unable to sustain projected consumption requirements. But with substantial discoveries in the Middle East and elsewhere, the U.S. has been able to import sufficient quantities to offset declining domestic production. U.S. oil production peaked in 1970 and has been declining ever since. Russia is currently the third largest oil producer, behind Saudi Arabia (which is number one) and the U.S., but Russian oil production peaked in 1984 and has also been in the process of declining ever since. Thus, in the future, both the U.S. and Russia will be increasingly dependent on the remaining oil reserves in the Middle East countries.

When the Middle East Arab oil embargo occurred in 1973, the price of oil increased by a factor of four (from $3 to about $12 per barrel). In addition to the psychological shock of not being able to obtain gasoline without great difficulty, a profound economic transition was set in motion by the oil price increases. Automobiles and houses began to get smaller and much more expensive, but something much more ominous began to happen; many of the energy-intensive steel and other smokestack industries went out of business. On the other hand, it was the loss of heavy industry, along with a worldwide recession, increases in Arab oil production, and the development of more energy-efficient automobiles, homes and appliances, which collectively resulted in the oil surplus which began in the early 1980s.

The bottom line on the current surplus of oil is that it is temporary. The initial gas lines during the 1970s were merely a warning sign that there is much more serious trouble ahead. Fortunately, the current global oil surplus has provided some breathing space to organize an energy transition without the panic reactions that usually result from waiting until the situation goes critical. This being the case, it has been unfortunate that the Reagan, Bush and Clinton administrations did not take advantage of the relatively low-cost oil reserves to implement a transition to renewable energy technologies and resources. Rather, these administrations have held the view that sufficient oil and coal will be found in the remaining wilderness and/or offshore areas in the U.S. and elsewhere. This, in turn, has only accelerated the exponential destruction of the remaining wilderness and wildlife habitats.

The Battery

The fossil fuels are like the battery in an automobile. There is enough stored energy in the battery to start the engine, but if someone seriously proposed to propel the vehicle on the energy stored in the battery alone, one is not going to get very far. To place this issue into perspective, consider Figure 3.5. Perhaps the most significant consideration is that in relative terms, the petroleum era will be a temporary one. Yet the global environmental damage that it is causing could be irreversible. It is not just a question that the survival of life on the Earth as we know it may lie in the balance.

There is also the day-to-day quality of life issues that have to do with living in polluted cities and being forced to breath air that causes disease and death. There is the hard reality that if there is a continued insistence on using the remaining fossil fuel reserves to provide fuel for industrial and vehicular engines, after a certain point, the fossil fuels will be quickly depleted. At that point, little -- if any -- economically recoverable energy reserves will be left to build the renewable resource infrastructure. That, in turn, means the cost of the transition will be many times greater. Most people, even reasonably well-educated people, are unaware of the significance of exponential growth. They are also probably not aware of the studies done by the U.S. Geological Survey, M.I.T., or the RAND Corporation, that clearly warn of the dangers of just continuing to consume the remaining fossil fuel reserves. They are also apparently not aware of the many studies that have documented the serious global environmental hazards that are associated with using fossil fuels for energy generation purposes.

Figure 3.5: The Battery
The oil era in a historical perspective
The illustration was prepared by Hubbert in 1962.

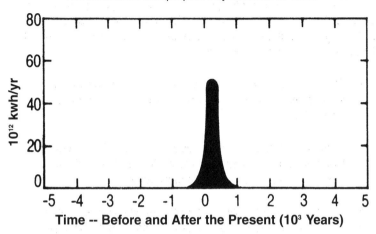

Time -- Before and After the Present (10³ Years)

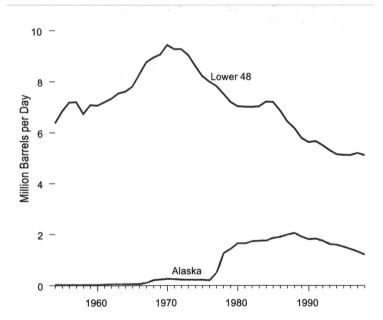

Figure : 3.6: U.S. Oil Production Levels
The data includes the lower 48 states and Alaskan crude oil production. Source: U.S. Energy Information Administration.

Natural Gas

The cost of electricity generated from natural gas is expected to only increase over time as the non-renewable reserves of natural gas are exponentially diminished. According to the U.S. Energy Information Administration:

- The total discoveries of natural gas reserves in 1998 were only 11,433 billion cubic feet, down 27 percent from 1997.

- New field discoveries in 1998 were less than half of the new field volume discovered in 1997, and 30 percent less than the prior 10-year average.

- Although the number of natural gas wells increased by 7 percent in 1998, the average of total discoveries per exploratory gas well actually fell by 32 percent. The Gulf of Mexico natural gas reserves also declined 5 percent in 1998.

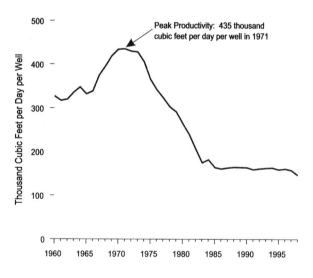

Figure 3.7: Natural Gas Well Productivity
Source: U.S. Energy Information Administration

U.S. natural gas production in 1998 was 19 trillion cubic feet, well below the record-high 21.7 trillion cubic feet in 1971. As Figure 3.7 indicates, natural gas well productivity peaked at 435 thousand cubic feet per day in 1971, and then fell steeply through the mid-1980s. Productivity in 1998 was only 146 thousand cubic feet per well per day. Although natural gas (which is mostly made up of hydrogen) is an important "bridge-fuel" to the hydrogen economy, the statistical data on the declining natural gas and crude oil reserves would suggest that the transition to renewable hydrogen technologies and resources should be pursued with "wartime" speed. It is also worth noting that proven crude oil reserves in the U.S. fell 7 percent in 1998, which was the largest decline in 53 years. Only 24 percent of oil production in 1998 was replaced by proved reserve additions. Figure 3.6 provides a historic trend for U.S. oil production.

Conclusions

Because fossil fuels are nonrenewable and are being exponentially consumed, they will be unable to sustain an expanding global industrial economy. Even if the fossil fuels were inexhaustible, however, their unacceptable environmental impact would still dictate that alternative energy technologies and resources be developed. To be unaware of such information is analogous to the captain of the *Titanic* not being aware that his ship was going to run into an iceberg. Because nuclear energy systems do not produce greenhouse gases or acid deposition, they are being suggested as the logical solution to the problems created by burning fossil fuels. As such, Chapter 4 will focus on nuclear energy systems, and examine their potential to be mass-produced for large-scale hydrogen production. However, as the data in Chapter 4 will show, nuclear technologies generate staggering environmental problems of their own that revolve around the radioactive wastes that have thus far proven to be unmanageable. Fortunately, there are a number of viable renewable energy technologies discussed in Chapter 6 that are capable of being mass-produced for displacing the use of fossil fuels, as well as nuclear energy systems.

Figure 4.1: Yucca Mountain
Potential home in Nevada for the high-level nuclear waste
generated in the U.S.

CHAPTER 4

NUCLEAR POWER

"The splitting of the atom has changed everything save our modes of thinking, and thus we drift toward unparalleled disaster."

Albert Einstein

Nuclear Fission

Nuclear-fission power plants were initially thought to be the answer to the diminishing fossil fuel reserves. Even though the enriched uranium fuel was also severely limited, it was assumed that a more advanced nuclear technology -- referred to as "breeders" -- would eventually be made commercially viable. Breeder reactors would actually be able to produce more radioactive "fuel," in the form of plutonium, than they consume. As a result, their plutonium fuel would be renewable. This was a great concept in theory, although biologists had continually warned that plutonium is one of the most toxic elements known, it is very difficult to handle, and it would remain deadly for over 250,000 years. In addition, if the "Plutonium Economy" was ever to become a reality, millions of tons of plutonium would have to be produced and shipped throughout the U.S. and other countries.

Figure 4.2: Operating Nuclear Power Plants in the U.S.
There are 104 nuclear reactors that are currently operating in the U.S.
The number of operating nuclear plants in the U.S. peaked in 1990
with 112 units. Source: U.S. Energy Information Administration, 1998.

**Status of All Ordered Units,
1953-1998**

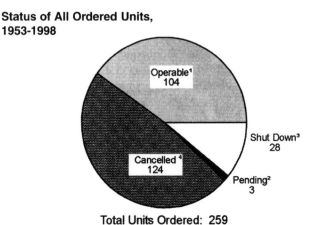

Total Units Ordered: 259

Operable Units,[1] 1957-1998

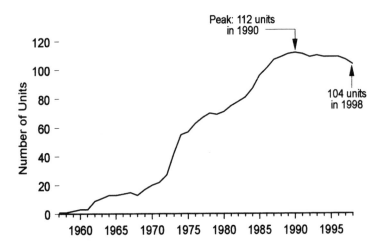

Figure 4.3: Status of All Ordered Nuclear Units
Source: U.S. Energy Information Administration, 1998.

In spite of these long-term and exceedingly difficult and dangerous environmental concerns, conventional nuclear reactors and their breeder offspring constituted America's primary energy strategy since the 1950s to resolve the diminishing fossil fuel problem. However, when the partial meltdown accident occurred in 1979 at one of the new nuclear reactors at Three Mile Island in Pennsylvania, both public and investor confidence in nuclear fission technologies were shattered. And although billions of taxpayer's dollars have been used to develop and promote nuclear energy systems, rather than solving the diminishing fossil fuel problem, nuclear technology has instead created an even more profound problem of its own -- radioactive waste.

The radioactive waste problem is unique for many reasons, but one of its most insidious aspects involves the fact that it is invisible to the human senses until disease or death occurs. When moisture is present radioactive isotopes spread in an ecosystem like red dye spreads in a glass of water, and the isotopes will remain toxic for anywhere from a few days to in excess of a million years, depending on the isotope. There is still no long-term storage plan for these waste products and many of the existing waste storage facilities are full and out of control in terms of their ability to prevent the radioactive wastes from leaking and spreading into the environment.

This "spreading" problem occurs because under irradiation, all materials change their nature as their atoms become unstable. The world of radioactivity is a world of continuous change. This is why the storage vessels which contain the radioactive wastes are only reliable for relatively short periods of time. Eventually, they too will become radioactive. In turn, this means the longer a nuclear reactor is operating, the more radioactive it becomes, That is the primary reason why any repair or maintenance of aging nuclear reactors is an extraordinarily hazardous task.

Radioactivity

The problem of radioactivity in the reactor and the surrounding containment building is in addition to the liquid, gaseous and solid waste generated by the uranium fuel rods. As the uranium undergoes fission, the uranium atoms split, and in so doing, release neutrons. Some of these neutrons split other uranium atoms, which, in turn, produce the radioactive waste products. The net result of this fission process is the generation of the intense heat that is used to generate steam for the production of electricity. The difference between a nuclear reactor and a nuclear weapon is measured by the number of neutrons that are released in the fission process over a given period of time. If only a limited number of neutrons are available for triggering the fission chain-reaction, the reaction can be controlled for energy production purposes. If too many neutrons are released, the chain-reaction will rapidly accelerate, resulting in a nuclear meltdown. To prevent this from happening, nuclear reactors have control rods and water circulation to regulate the fission process by absorbing the extra neutrons.

However, during normal operations, some of the neutrons that are released pass beyond the uranium and into the steel structures which hold both the fuel assemblies and the cooling water that flows between them. Other neutrons penetrate the massive concrete shielding outside the steel reactor vessel. These neutrons are absorbed by the atoms of iron, nickel and other elements that make up the steel, water and concrete. When atoms absorb neutrons, they are rendered "unstable" (i.e., radioactive) for various lengths of time. In the case of nickel-59, which has a half-life of 80,000 years, it will need to be shielded from humans for about a million years.

Decommissioning

Decommissioning is the term used to describe what happens to a nuclear reactor and its related systems once their theoretical useful life of 30 or 40 years has been completed. This "back end" of the nuclear fuel cycle is generally not discussed because no one really knows how such a difficult task is to be accomplished, or what it will ultimately cost over hundreds of thousands of years. Utilities and waste processing companies are not concerned about the long-term waste storage costs because in most cases, they have no long-term legal or financial responsibility to manage the radioactive wastes. That responsibility is given to the U.S. taxpayers of the future. But if the experience of attempting to decommission a nuclear fuel recycling plant at West Valley, New York is any indication of what is ahead, there are plenty of reasons to be concerned.

The $32.5 million West Valley plant, located 30 miles southeast of Buffalo, was officially opened with much fanfare in June of 1963, although it did not actually begin to reprocess nuclear wastes until 1966. But it was only to operate for 6 years before its operator, Nuclear Fuel Services (NFS), a subsidiary of W.R. Grace's Davison Chemical Company, abandoned the facility. Left behind were 2 million cubic feet of buried radioactive trash and 600,000 gallons of highly radioactive liquid waste that is now seeping into the Cattaraugus Creek, which flows into Lake Erie, from which the city of Buffalo obtains its drinking water. The cost of cleanup is estimated to be at least $1 billion, assuming such a cleanup is even possible.

The West Valley plant was the world's first commercial nuclear-waste facility that would be able to take the spent fuel from nuclear power plants and reprocess it for renewed use. This recovery and reprocessing of spent fuel rods is an important step in the nuclear fuel cycle because only about 1 to 2 percent of the nuclear fuel is initially consumed in most commercial reactors, and uranium is not a renewable resource. This is particularly true of the reactors in the U.S. that require energy intensive, highly enriched uranium 239. Unfortunately, the problems associated with trying to handle highly radioactive wastes proved to be overwhelming to the workers and management at the West Valley facility. An extensive investigative report on the West Valley plant titled "Too Hot to Handle," that was undertaken by *The New York Times*, concluded the following:

> "... It is the story of technocrats who assured and reassured the public that nuclear recycling was safe and that a thoughtfully engineered fail-safe system would minimize the hazards of any accidents that might possibly occur -- without making it clear that their assurances were based on extrapolations from premises rooted in probabilities and anchored in uncertainty. It is the story of company officials who repeated such assurances even after scores of incidents -- known only inside the company and to a few governmental inspectors -- had made it clear that leakage of radioactivity in the plant was reaching dangerous levels."

Radiation at the West Valley facility was rapidly spreading, and the costs of operation were dramatically increasing. Finally, in 1975, NFS announced that it would have to spend at least $600 million to make the facility manageable, which was nearly twenty times the initial capital cost of the plant. This, plus the fact that operating costs had increased by some 4,300 percent since the plant began operation, left NFS with little other choice but to shut the facility down, and let the taxpayers of New York figure out what to do with the billion dollar waste storage problem that was left behind.

Nuclear Waste Storage

As it turns out, the West Valley facility in New York is merely the tip of the nuclear waste iceberg. The Department of Energy spent years and $700 million to build the nation's first permanent waste storage facility deep in salt beds near Carlsbad, New Mexico. But only after the facility was completed in 1988 did the DOE officials discover that water had somehow leaked into the salt caverns. As a result, the Carlsbad facility has been put on hold, but the nuclear waste problem in the U.S. is so critical that Idaho Governor Cecil Andrus "declared war on the Department of Energy" in October of 1988. Under the governor's orders, state police halted any further shipments of nuclear waste into the State of Idaho.

The nuclear waste problem has been around for decades, and in spite of the fact that thousands of brilliant scientists have been working on the problem, no one in any country has come up with an acceptable long-term solution. It is really a question of deciding which of the existing imperfect alternatives will minimize the hazards of storing the wastes for such long periods of time. As the general public learns more about the waste problems, they become more and more convinced that they don't want the toxic dump in their state. Given this bleak outlook, the U.S. Congress decided just hours before the Christmas adjournment in 1987 that a place called Yucca Mountain, located in a remote nuclear test site in Nevada, was to be the nation's final resting place for high-level radioactive waste. Not only was the area already radioactive, but Nevada simply had less political influence in Congress than any other state.

However, there are a number of serious environmental problems associated with the Yucca Mountain site. First, scientists point out that the Yucca Mountain area is in a geologically unstable region with active volcanoes and earthquakes. Second, and perhaps more worrisome, is the fact that the containers for the nuclear waste will not be cooled in water pools to dissipate the intense heat. As a result, the containers of the waste will be lucky to last for even a decade. After that, it must be the mountain itself that contains the intense heat of the waste products, and therein lies the problem. According to the Draft Environmental Impact Statement for a Geologic Repository for the Disposal of Spent Nuclear Fuel and High-Level Radioactive Waste at Yucca Mountain, published by DOE in July of 1999:

"A substantial amount of uncertainty is associated with estimates of long-term repository performance."

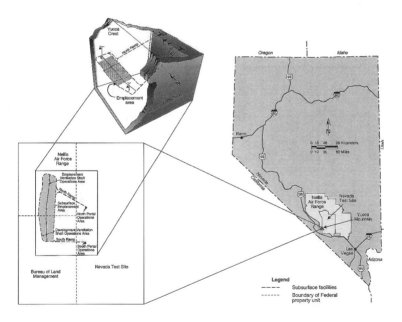

Figure 4.4: Location of the Yucca Mountain Site
Located on the edge of the Nevada Nuclear Test Site, the Yucca Mountain facility will likely become the nation's long-term high-level nuclear waste storage facility. The image was provided courtesy of the U.S. Department of Energy.

Thermal Load

The heat generated by spent nuclear fuel rods and high-level radioactive waste packages creates a "thermal load," which could affect the long-term performance of the nuclear waste repository (i.e., the ability of the engineered and natural barrier systems to isolate the radioactive wastes from the natural environment.) The thermal load can also potentially affect the short-term repository attributes, including the amount of surface area that will be required for construction and operations, the number of workers, and the electrical requirements. Most of the thermal load from the radioactive wastes comes from the spent fuel rods from commercial nuclear reactors. In commercial nuclear reactors, the heat from the spent fuel rods is dissipated by placing used rods under water. The problem with this approach is that these temporary storage facilities, that are located outside of the protective containment structure, are rapidly filling up.

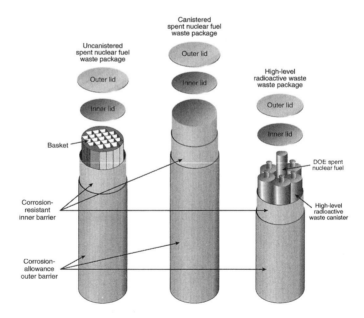

Figure 4.5: Potential Waste Package Designs
According to DOE, the spent nuclear fuel and high-level nuclear waste packages would range in size from 12 to 20 feet long and 4.1 to 6.6 feet in diameter. They would weigh from 77,000 to over 183,000 pounds.

Figure 4.6: Nuclear Waste Storage Containers
An artist's conception of a nuclear waste package in what is referred to as an "emplacement drift." According to DOE investigators, the heat generated by the nuclear waste packages (which will not be cooled with water) could affect the geochemistry, hydrology, and mechanical stability of the emplacement drifts, which in turn could influence groundwater flow and the transport of radionuclides from the engineered and natural barrier systems to the environment. The image was provided by DOE.

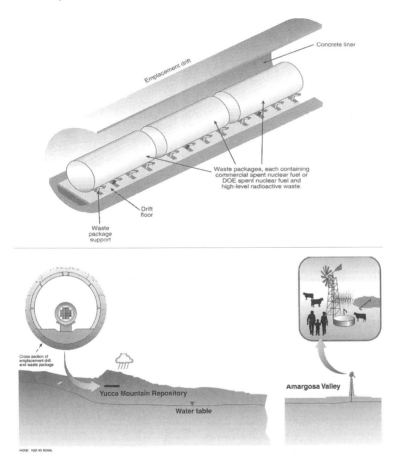

Figure 4.7: Potential Movement of Radioactive Wastes
According to DOE, the intense heat that is generated by spent nuclear fuel and high-level radioactive waste (the thermal load) could adversely affect the long-term performance of the repository (i.e., the ability of the engineered and natural barrier systems to isolate the radioactive wastes products from the accessible environment for over one million years.)

As part of the analysis of the Yucca Mountain site, scientists from the U.S. Geological Survey drilled two small shafts in Yucca Mountain. What they found was that the mountain "breathes." In the winter, warm air flows out of the shafts from the center of the mountain. In the summer, the pattern is reversed and the air is sucked into the mountain. In a memorandum, 17 scientists from the USGS stated that this airflow could release radionuclides into the atmosphere. If that were not bad enough, in 1989, officials from the U.S. Department of Energy announced that in spite of the critical nature of the nuclear weapons program, all three of the nuclear weapons facilities had to be shutdown due to massive technical problems associated with dealing with the radioactive waste. Department of Energy officials have now admitted that extensive releases of radiation had often contaminated workers and large numbers of U.S. civilians who happened to be living within the proximity of these facilities. The general public was deliberately not informed of these radiation leaks for decades because of "national security" considerations. As one federal official put it, ". . .the emphasis was not on safety, but on production."

It has been estimated that it will take anywhere from 100 to 150 billion dollars to clean up the existing nuclear weapons facilities, assuming such a clean-up is even possible. In reality, such clean-ups have typically meant collecting and moving as much of the waste material as possible to some other location, which ultimately only creates another waste problem in the future that will eventually have to be dealt with. As a practical matter, efforts at removing radioactive isotopes from existing areas have up until now proven to be only marginally effective; and ultimately, the radioactive waste materials are simply removed from a site and sent to another site, such as Yucca Mountain.

These unresolved technical problems have led to the concept of "National Sacrifice Zones," which implies sealing-off the contaminated areas from the members of the general public for literally hundreds of centuries. Such a policy in and of itself poses a wide-range of interesting social and ethical problems. For example, imagine trying to figure out how to make a sign or symbol that could explain the deadly nature of the radioactive waste that could still be understood by our descendants thousands of years into the future. One can only wonder what the future generations will think of this generation for leaving them such seemingly unresolvable and deadly problems.

Questions of Safety

Proponents of nuclear power like to point out that it is highly unlikely for a nuclear fission power plant to explode like a nuclear bomb, which is true. They rarely explain, however, that given the amount of radiation in a utility-scale 1,000 MW reactor, a serious loss of coolant accident could result in a meltdown condition that could be as dangerous as an outright nuclear explosion. This is because the amount of radiation in a typical 1,000 MW reactor is more than a thousand times that which was produced by the atomic bomb that was dropped on Hiroshima in 1945. Moreover, in a meltdown scenario of a conventional nuclear reactor, the massive amounts of deadly radiation in the reactor vessel would virtually all be released at groundlevel. This is in contrast to a nuclear explosion that would disperse much or most of the radioactive particles high into the Earth's atmosphere. This is essentially what happened when the Chernobyl accident occurred in Russia in 1986. The resulting radioactive cloud then drifted over Western Europe and contaminated large areas of a number of countries.

Although the Soviet government at the time initially assured those living in the area of the Chernobyl accident that the medical impact would not be significant, three years after the accident cancer rates doubled among residents of contaminated farm regions, and calves and other farm animals have been born without heads, limbs, eyes or ribs. *Moscow News* has reported that more than half of the children in the Narodiehsky region of the Ukraine have illnesses of the thyroid gland, which exposure to radiation can cause. Russian officials now admit that they "drastically under-estimated" the health problems caused by the Chernobyl reactor meltdown, explosion and fire.

Because of the serious health issues that are associated with radioactivity, once a nuclear reactor becomes operational, technicians are not able to clean and maintain the critical interior components of the reactor vessel due to the intense level of radiation. This is in contrast to the normal procedure in fossil-fueled power plants where maintenance personnel are able to shut the plant down periodically to clean and inspect all of the critical interior components. This regular maintenance schedule is a major reason why there are so few accidents or failures in fossil-fuel power plants, in contrast to nuclear fission facilities, whose unreliability has become one of their most dependable features.

In contrast to the civilian nuclear reactors that have been plagued by accidents and shutdowns, the U.S. Navy has established a respectable performance record with its nuclear surface ships and submarines. There are, however, distinct differences in the size and quality of the nuclear systems used by the U.S. Navy, in contrast to the much larger reactors engineered for commercial power production. The most obvious differences involve the reactor size, design configuration, and cost per installed kilowatt (kW). The reactors manufactured for the U.S. Navy are all designed to be relatively small; they are all based on a similar design concept; and they were initially about five times more expensive to construct (in terms of cost per kW) than their civilian counterparts.

William E. Heronemus, who was then professor of civil engineering at the University of Massachusetts, supervised the construction of nuclear submarines for Admiral Hyman Rickover. In 1973, Heronemus testified before a nuclear licensing hearing that it would cost roughly $2,400 per kW to build one of the U.S. Navy's nuclear reactors, compared with $400 per kW for the commercial plant that was under review. (Note that at present, capital costs for commercial reactors are closer to the $2,400 figure.) Heronemus also testified that he would have refused to approve the wiring and piping for the Navy that had been accepted and incorporated in the commercial nuclear facility.

Advanced Reactor Considerations

According to Charles E. Till, a nuclear physicist at the Argonne National Laboratory in Illinois, a new generation nuclear fission reactor, referred to as an "Integral Fast Reactor," has been under development by the U.S. Department of Energy for several years. This new liquid sodium-cooled reactor configuration is expected to be safer, minimize corrosion and be more efficient (i.e., it should be able to use 15 to 20 percent of the uranium fuel instead of the 1 to 2 percent with current reactor designs), and in the process, generate less radioactive wastes than the existing generation of light-water reactors in use in the U.S. In addition, according to Jerry Griffith, an associate deputy assistant secretary for reactor systems development for the DOE, the Integral Fast Reactor is considered especially crucial to the future of nuclear power because "it is the best technology for breeding plutonium." As the world's uranium reserves become scarce, plutonium will be needed as a substitute nuclear fuel. Thus, the Integral Fast Reactor is the key to a nuclear future and a "plutonium economy." However, there are three primary concerns with the advanced Integral Fast Reactor:

1. Nuclear physicists have had a relatively poor track record in predicting the engineering outcome of many theoretical calculations, and thus far, not even a prototype of the Integral Fast Reactor has been tested.

2. Liquid sodium is an extremely volatile substance that will burst into flames if it comes into contact with either air or water. Two liquid sodium-cooled nuclear breeder plants in the U.S. were totally destroyed by liquid sodium fires.

3. Plutonium is an exceedingly difficult and dangerous material to handle. It is one of the most toxic elements known. Plutonium is 35,000 times more lethal than cyanide poison by weight. There is an equivalent of 20 million mortal doses in only 5 grams of plutonium, which is the weight of a 5-cent coin, and it will remain dangerous for hundreds of thousands of years. This long-term toxicity of plutonium, in and of itself, creates significant moral and ethical questions about producing such long-lived toxic substances that are invisible to human senses.

It is important to realize that in order to produce enough energy to displace fossil fuel resources, which now account for roughly 90 percent of the industrial world's energy supply, immense quantities of plutonium would have to be created. And given the extremely poor track-record of containing or properly disposing of the plutonium and other radioactive wastes that have already been created, the construction of the thousands of plutonium-fueled reactors that would be necessary for the "plutonium economy" to be viable would hardly seem to be an acceptable alternative. These complex, long-term problems underscore the importance of evaluating the renewable energy options that do not pose such long-term unknown environmental and economic risks.

Nuclear Fusion

Nuclear fusion reactors, in contrast to nuclear fission reactors, do not split uranium atoms. They are intended to fuse hydrogen atoms in a process similar to that which occurs in the Sun and other stars. Although fusion physics is a common occurrence in stars, it is well to remember that no biological organisms are able to live in such high-temperature environments. Billions of dollars have already been spent on this highest of high-tech energy technologies, which has been under development for decades by governments in the United States, Japan, France, Germany, the former Soviet Union and other European countries. However, it has been as a result of this research that at least some fusion reactor advocates are now questioning whether such high-temperature (over 100 million degrees F) fusion energy systems will ever play any role in energy production during the next 50 to 100 years.

According to a nuclear fusion article by John Horgan published in the February 1989 issue of *Scientific American,* the initial hopes of having small, safe high-temperature fusion reactors burning cheap, abundant fuel have all but disappeared. It is now estimated that exotic and expensive fuels will be required, they will produce significant quantities of radioactive waste, and even the smallest fusion reactor would be comparable in size and complexity to the largest of today's fission reactors. Such high-temperature fusion technologies face staggering technical problems, and billions of additional dollars will be needed just to build a prototype. Even if the prototype plant actually works from a technical perspective, the really important question is whether nuclear fusion systems will ever be cost effective. As such, it is irrational to predicate a nation's energy policy on such high-risk and unproven technologies. In spite of this reality, the U.S. Congress continues to spend over $300 million annually on fusion research and development.

Cold Fusion

On Thursday, March 23, 1989, two scientists, Stanley Pons, Chairman of the Department of Chemistry at the University of Utah, and his colleague, Martin Fleischmann, professor of electrochemistry at the University of Southampton, England, stunned the scientific world when they held a national press conference to announce the results of their "cold fusion" experiments. Their announcement indicated they had not only succeeded in generating a fusion reaction that actually produced more energy than the reaction consumed, but they had accomplished it at room temperature, in a simple table-top test tube apparatus in their kitchen. They did not indicate what radioactive isotopes were produced as a result of their room temperature reaction. The two scientists indicated that they had been working on their cold fusion process for more than 5 years, and they felt that commercial reactors based on the new low-temperature fusion process could be in operation in about twenty years.

While the news media covered the story worldwide, most scientists were highly skeptical of such claims, and they were troubled that such an announcement came at a press conference rather than having a paper published through the traditional peer-review process. As the world now knows, the cold fusion process did not work. It was just another claim that turned out not to be true.

Nuclear Economics

The true cost of nuclear power has been confused by the quasi-public nature of the research and development. U.S. taxpayers are financially responsible for the "back-end" of the nuclear fuel cycle, which includes covering any costs not met by the utility for waste disposal and decommissioning. Billions of taxpayer's dollars have also been spent for the "front-end" of nuclear research and development. These costs are not included in most nuclear cost totals. They include the construction and operation of the three U.S. uranium fuel enrichment facilities that are at present shut down due to the extensive problems with respect to radiation spreading. When all three of these enrichment facilities were operating at full capacity, their electrical requirements were actually about the same as those used by the entire country of Australia. Other excluded costs of nuclear power include the high-level of federal regulation that is required, as well as the long-term waste disposal and the numerous health costs that are associated with people being exposed to radiation.

To comprehend the nuclear issue, it is necessary to put time into perspective. If toxic wastes, which will be deadly for 200,000 or 500,000 years, are generated, is it possible for anyone to comprehend the actual environmental or economic costs? The very first civilized groups of people in the Middle East appeared only about 8,000 to 10,000 years ago. How is it, then, that one generation could, or should, assume the right to create insidious radioactive hazards that will remain deadly for hundreds of thousands of years? How is it that we have allowed ourselves to do such things that cannot be comprehended or calculated in terms of cost or human death and disease?

The nuclear and other toxic waste problems are global in nature and have clearly transcended the capitalist or communist ideologies, as both political systems have developed nuclear technologies. It is interesting to note that both the U.S. and the former Soviet Union -- as well as most other countries -- have up to now shown a complete disregard for the "human rights" of future generations. It seems difficult to imagine how so many "civilized" nations could have allowed the production of such deadly and long-lived radioactive wastes to occur. Even more difficult to understand is how the citizens in the same countries can silently let the production of such long-lived toxic wastes continue. If the ovens of the prison camps in Nazi Germany are now viewed as a moral outrage, one can only wonder how future generations will view the actions of the present generation.

Because of the development and use of nuclear energy systems, the human community is now faced with an awesome array of problems, and unlike most other environmental problems, acceptable solutions for the disposal of radioactive wastes are as yet unknown. This hard reality is underscored by the fact that after more than 30 years of concentrated effort by a wide-range of distinguished scientists from around the world, no one has yet demonstrated a genuine solution to the long-term radioactive waste problem. Indeed, in a detailed study done by the Jet Propulsion Laboratory of the California Institute of Technology for the President's Office of Science and Technology Policy, the following conclusions were observed:

> *"The problems of high-level nuclear waste management are so complex and have so many ramifications that no one person or group of persons can possibly have all the answers. The results of this study indicate that the U.S. program for high-level waste management has significant gaps and inconsistencies."*

As a result, it would seem that far from solving the rapidly diminishing fossil fuel problem, the nuclear power industry has only succeeded in creating a whole new range of technical and long-term environmental problems that will be inherited by our children and their children for thousands of generations into the future. Barry Commoner made a similar observation in his book, *The Closing Circle*:

> *"Our experience with nuclear power tells us that modern technology has achieved a scale and intensity that begins to match that of the global system in which we live. We cannot wield this power without deeply intruding on the delicate environmental fabric that supports us. It warns us that our capability to intrude on the environment far outstrips our knowledge of the consequences."*

Placing such an ominous issue as radioactivity into an evolutionary perspective is not easy. Kenneth and David Brower have expressed it as well as anyone in an article, "Miracle Earth" that was published in *Omni* magazine. In their article, they point out that when a beta particle (a high-speed electron emitted from the nucleus of a radioactive atom) strikes living tissue, "it rips negatively charged electrons from the tissue's atoms, leaving positively charged ions in its wake." These liberated electrons, in turn, ionize other atoms in a cascading effect, which proceeds to tear apart tens of thousands of highly ordered biological molecules that serve as the structure of living cells. "The passage of such a particle leaves the city of the cell in ruins. Alpha, gamma, and x-rays all have this effect on biological molecules. Their entry hole is small, but their exit hole is spectacular." Kenneth and David Brower go on to describe the delicate balance that is held between life and radiation:

> *"Life is adrift in a sea of radiation, as the makers of the artificial kind are apt to point out... Radiation seeps up from springs on Earth and flows in rivers from space. The sun sends out a steady stream of particles freshened occasionally by solar storms. Yet on Earth, life has found something like a back-water. Here life is protected from the full force of the cosmic stream by the planet's atmosphere; here life tolerates the weaker radiations*

from the planetary crust... But the margin within which life operates is small, and the tolerances are fine. There is no question who the enemy is, our first and oldest. If the universe is hostile to life, there is no better expression of that hostility than the radioactive particle. The Four Horseman are secondary enemies, at a less basic level of organization. The particle attacks us fundamentally by disordering the atoms of which we are made. It strikes at and scatters the miraculous principle that distinguishes us from dust."

Conclusions

Because nuclear energy systems do not produce greenhouse gases or acid deposition, they are being offered as the logical solution to the problems created by burning fossil fuels. As this chapter has documented, however, nuclear fission technologies generate staggering environmental problems of their own that revolve around the radioactive wastes that have thus far proven to be unmanageable. Moreover, if nuclear power plants were to effectively replace the burning of fossil fuels, about 10,000 gigawatt-scale reactors would have to be built at a cost of over $10 trillion. Since the existing uranium reserves will barely be able to keep the 104 existing reactors in the U.S. operating much beyond the year 2000, a whole new type of untested breeder nuclear technology would have to be rapidly developed. For all of these reasons, if nuclear systems were the only energy alternatives to the burning of fossil fuels, there would be little reason to be optimistic about the future of the human community.

Fortunately, there are several viable renewable solar-hydrogen technologies discussed in Chapters 5 and 6, and the renewable resources discussed in Chapter 7, that are capable of displacing the use of both fossil and nuclear fueled energy systems. While the political aspects of reordering national priorities are very real, most elected officials determine their issue priorities by finding out what the majority of their constituents think is important. Although elected officials are often criticized for tailoring their views to the whims of public opinion, they do have a responsibility to be aware of how the majority of the people they represent feel about issues.

This essentially means it is necessary for the majority of voting citizens to be informed about the global energy and environmental problems which are continuing to erode exponentially, and that an industrial transition to renewable energy resources represents a fundamental solution to these problems. Once the majority of the voting public understands these key points, a bipartisan political mandate to reorder national priorities around a transition to renewable energy resources can be initiated.

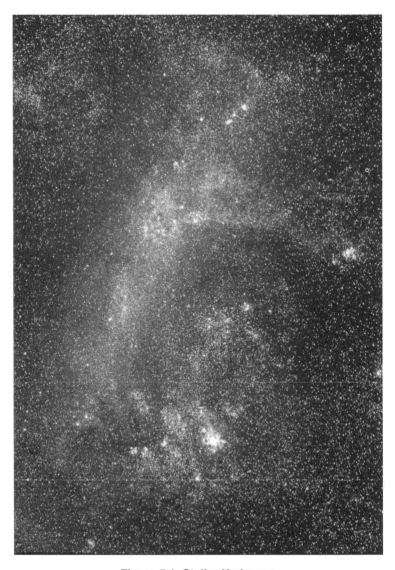

Figure 5.1: Stellar Hydrogen
Giant molecular-cloud complexes, consisting almost entirely of hydrogen, are the most massive objects within galaxies. Gravity is the primal force that eventually causes the hydrogen to compress until it can fuse into the heavier chemical elements. This being the case, hydrogen is not only the primary fuel for the Sun and other stars, it is also the primary building block of matter in the known universe. The above photograph was provided courtesy of the National Optical Astronomy Observatories.

CHAPTER 5

HYDROGEN

In the Beginning

Since the very beginnings of recorded history, there have been people who have revered the Sun as the source of life. This should not be surprising. It is fairly obvious that without the energy emitted by the Sun, life as we know it could never have evolved on the Earth. Few people, however, realize that the primary fuel for the Sun and other stars is hydrogen. Although the primal force that causes the Sun and other stars to burn is gravity, the primary fuel is hydrogen, and in the case of our Sun, it consumes about 400 million tons of hydrogen every second. As this primordial hydrogen is burned, or more accurately, "fused" into helium, photons of electromagnetic energy are released. The photons are eventually filtered through the Earth's atmosphere as solar energy. Thus solar energy is the result of a nuclear fusion process (not to be confused with the nuclear fission process of conventional nuclear reactors), without which there would not only be no life; there would be no fossil fuels; no wind; not even any uranium.

Given the need to make a fundamental transition from the nonrenewable fossil fuel and nuclear energy systems, it is only reasonable to consider the development of renewable energy technologies that could utilize the inexhaustible energy of the Sun. Because it is reasonable to assume that solar energy will someday serve as the primary energy source, it is important to understand the fundamental relationship that exists between solar energy and hydrogen.

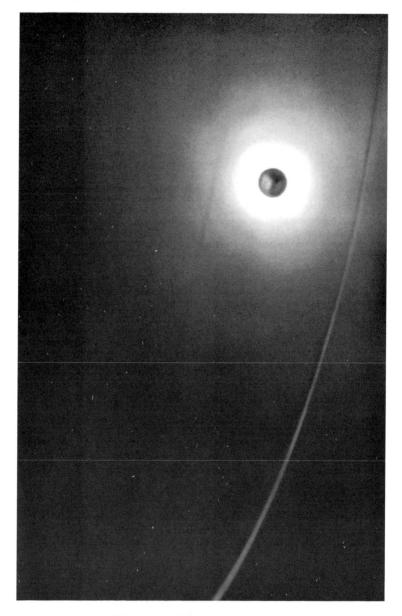

Figure 5.2: The Hydrogen Atom
The hydrogen atom is made up of a single proton and a single electron. These subatomic components were created within the first three
seconds of the "Big Bang" that occurred some 15 billion years ago .

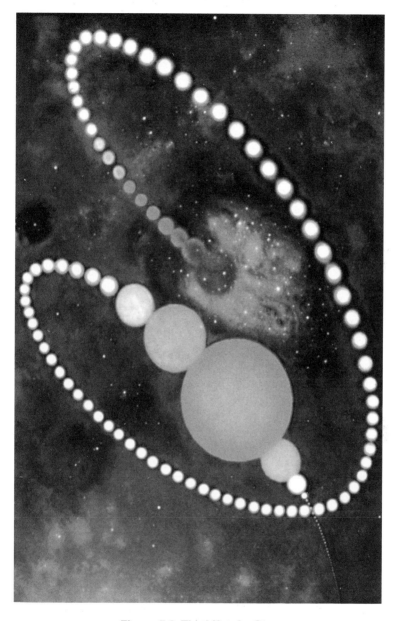

Figure 5.3: The Life of a Star
Gravity is the primal force that caused the hydrogen atoms in the universe to condense into stars, which then burn the hydrogen fuel and resulting heavier elements until they are exhausted.

Primordial Hydrogen

Modern physics provides a fairly clear picture of the origin of matter in the known universe, at least with respect to the development of the proton and the electron, which are the basic components of the hydrogen atom. Hydrogen atoms, in turn, are the basic building blocks of all of the other 91 chemically distinct atoms that occur naturally. The atomic number of an atom is equal to the number of protons (i.e., hydrogen nuclei) -- or electrons it contains. Thus hydrogen, with one proton and one electron, has an atomic number of one. Carbon, on the other hand, has six protons and electrons, and therefore has an atomic number of six. The proton has a positive electrical charge, and the electron has a negative charge, which explains their natural affinity for each other. These basic elements are believed to have been formed within the very first second of the origin of the universe when the "Big Bang" occurred some 15 billion years ago.

Were it not for gravity, which is the basic force that causes every particle of matter in the universe to be attracted to every other particle, the hydrogen atoms and other subatomic particles would have continued to fly away from each other from the initial force of the big bang. It was gravity, therefore, that was the primal force that caused the interstellar particles to concentrate in larger and larger masses. As the mass increases, so does the force of gravity, and eventually, the forces and pressures are sufficient for the interstellar clouds of hydrogen to collapse. When this collapse occurs, the hydrogen and other particles collide, and as a result, high enough temperatures (45 million degrees Fahrenheit) and pressures are created for the hydrogen to fuse into helium, at which point a star is born.

As the stars burn their supply of hydrogen, four hydrogen nuclei are fused into a heavier helium nucleus. The helium forms a dense, hotter core, and as the star consumes most of its primary hydrogen, it begins to burn -- or fuse -- the helium, converting it first to carbon and eventually to oxygen. Thus, a star is like the legendary Phoenix bird, destined to rise for a time from its own ashes of heavier elements. The more massive stars achieve higher central temperatures and pressures in their late evolutionary stages. As a result, when their helium is consumed, they are able to fuse the carbon and oxygen into still heavier atoms of neon and magnesium, and eventually silver and gold.

Carl Sagan summarized this process in his book *Cosmos:*

"All the elements of the Earth except hydrogen and some helium have been cooked by a kind of stellar alchemy billions of years ago in stars...The nitrogen in our DNA, the calcium in our teeth, the iron in our blood, the carbon in our apple pies were all made in the interior of collapsing stars. We are made of starstuff."

It should be clear from this perspective of physics and chemistry, that the relationship between solar energy and hydrogen is inseparable, in the sense that without hydrogen, there would be no Sun. In a somewhat similar context, the evolution of life itself on the primitive Earth was also a result of the dynamic interactions of hydrogen and solar energy.

The Nanobes & Microbes

At present, no one knows exactly how the first living organisms evolved out of nonliving matter. What is known is that atoms of hydrogen, carbon, oxygen and nitrogen eventually did evolve into an electrochemical structure of amino acids. These amino acids were, in turn, assembled into the highly complex architecture that makes up proteins, which includes the enzymes that are so critical to metabolism. Molecular biologists have been able to show that these building blocks of life may have been the first, and the most basic molecules of living organisms. Since microbes such as bacteria, fungi and viruses operate on the micrometer scale (i.e., one-millionth of a meter), they have been referred to as microorganisms, or microbes. But at the heart of the microbes are the enzymes and other proteins that operate on the nanometer scale (i.e., one-billionth of a meter). This being the case, these protein-scale organisms may be thought of as "nanoorganisms," or "nanobes," and given their seemingly incredible array of activities, the nanobes could be viewed as a highly-advanced civilization that is more than 3.5 billion years old. Although the human brain and nervous system is a marvel of sophistication, it is literally the end product of billions of years of nanobial evolution, and it is directly operated and maintained by a current generation of nanobes that are at the heart of life itself.

Figure 5.4: The Evolution of Life on Earth
The age of microorganisms dominates the time span of biological
evolution. The illustration is from "Archaebacteria," by Carl R. Woese.
Copyright © June 1981 by *Scientific American, Inc.* All rights reserved.

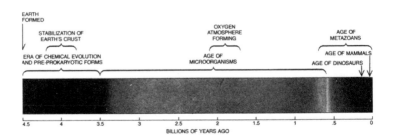

It would appear that the nanobes are able to manufacture
microbes, human beings and other animals in the same way that
humans make aircraft carriers, and just as an aircraft carrier is
ultimately controlled by a group of specialized officers on the
bridge, there are specialized nanobes that occupy and regulate
the human command and control centers within the central nervous system. Daniel L. Alkon, chief of the laboratory of molecular and cellular neurobiology at the National Institute of
Neurological and Communicative Disorders and Stroke, has specialized in understanding the molecular mechanisms of memory.
In a paper "Memory Storage and Neural Systems," published in
the July 1989 issue of *Scientific American*, Alkon explained that
when individual neurons (i.e., brain cells) are in the process of
learning, the flow of potassium ions through channels in the
membranes is sharply reduced. The potassium-ion flow is what
enables nerve cells to conduct electrical impulses, and when it is
decreased, the cell becomes excited and electrical impulses can
be triggered more readily. What is significant is that it is the
nanobial enzymes and other proteins that appear to regulate this
critical ion flow. Other nanobes are responsible for programming
and maintaining the DNA biocomputer, which stores the memory
of amino acid sequences that are the genetic code for all the different types of nanobial proteins that exist within living organisms.

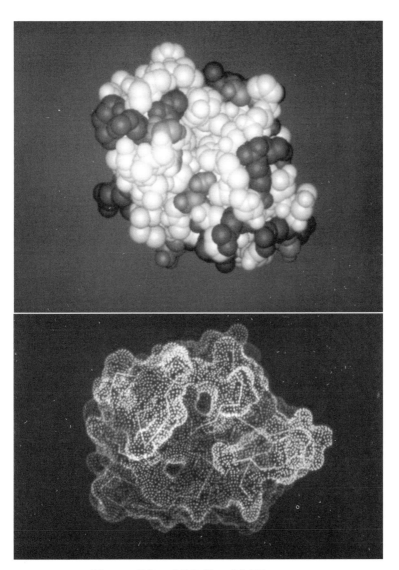

Figures 5.5 and 5.6: Nanobial Enzymes
Enzymes and other proteins are at the heart of metabolism. Both images are of the enzyme subtilisin. The top image is of the molecular structure of atoms (excluding hydrogen), whereas the bottom image is a dot pattern of the enzyme's surface that was generated by a Cray super-computer. Other enzymes have completely different structures and appearances. The computer graphics were prepared by Arthur J. Olson, Ph.D., Copyright © 1985 Research Institute of Scripps Clinic, California.

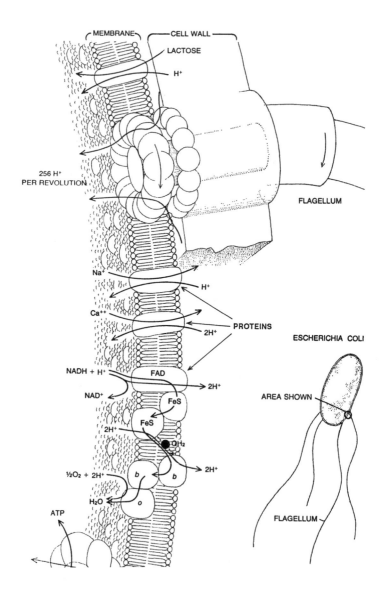

Figure 5.7: A Microbial Hydrogen-Fueled Engine
Note that the flagellum (i.e., the tail) of a typical bacteria is actually a molecular-scale rotary engine that requires 256 hydrogen nuclei (protons) per revolution. The above image is from the article "How Cells Make ATP," by Peter C. Hinkle and Richard E. McCarty. Copyright © March 1978 by *Scientific American, Inc.* All rights reserved.

Nanobial proteins are the actual building blocks of life, and depending on their configuration, they can become the structural material from which living tissue is made; they can become the hormones that regulate chemical behavior; or they can become the enzymes that mediate virtually all of the biochemical reactions in living organisms. Thus nanobes represent a critical link in the origin of life; they stand at the threshold of life where chemistry becomes biology. They are, in a biological sense, our creators. Although nanobial enzyme reactions were first observed in the 1830s, little was known about their evolution prior to the 1970s. This was because the surface imprints of bacterial fossils provided little information about their origin or what their internal molecular structure could have been like. However, with the recent advances in molecular biology, scientists are no longer limited to just analyzing the geological fossil record. It is now known that by carefully analyzing the amino acid sequences within a cell's proteins and the nucleotide sequences of its nucleic acids (i.e., DNA and RNA), it is possible to provide a remarkably accurate picture of an organism's genetic and molecular origin. This living record provides far more information than geologic fossils because it provides a detailed picture of the evolution of molecular structures. Moreover, the living molecules reach back to a time long before the oldest fossils, to a period when the common ancestor of all life existed.

The primary objective and function of the initial primordial organisms was -- and still is -- to extract energy from other molecules in their immediate environment. This was generally accomplished by rearranging the hydrogen atoms that bonded the various molecules. This movement of hydrogen atoms has been referred to as "transhydrogenation" or "fermentation," and such primordial metabolic processes were then, and are now, the essence of what the biochemistry of life is all about. Thus living organisms have been successfully using the hydrogen energy system for several billion years. The nanotechnology of transhydrogenation enabled the initial nanobe and microbe populations to increase exponentially, but they initially developed their hydrogen energy process around the nonrenewable hydrocarbon molecules they found in the primordial soup. This process went on for about 500 million years, at which point (roughly three billion years ago), their ever-increasing consumption of their nonrenewable resources began to exhaust their existing energy reserves.

Photosynthesis

In the same way that the nonrenewable primordial soup was being exponentially consumed by the Earth's early life forms, the current global reserves of nonrenewable fossil fuels are being exponentially consumed by the human population. Because the primordial soup was rapidly being exhausted, the nanobes were forced to modify their remarkably sophisticated molecular machines to utilize the vast and inexhaustible input of solar energy. The new molecule they developed was chlorophyll, and the relatively complex process it involves is referred to as photosynthesis. This innovative nanobial technology was one of the crucial links that allowed energy from the Sun to be incorporated into living, biological systems.

The eventual result of this nanobial reindustrialization effort, which allowed the nanobes to extract hydrogen from water with solar energy, was the development of green plants. The fabrication of photosynthetic molecules represented a major technological breakthrough that allowed the nanobes to make a successful transition to renewable solar-hydrogen resources. In addition, the problems of energy storage and transportation were also solved because the green plants could use the hydrogen they liberated from water to produce carbohydrates and fats. As a result, the umbilical link with the primordial soup of non-renewable organic molecules was finally severed, and the nanobe's foreseeable future was secure.

The scale and pace of evolution has increased enormously over the three billion years since the nanobes solved their energy crisis. But the basic molecular technology that they developed is still used today by their descendants, which occupy every living organism, including of course, the nucleus and cells within each and every human being. Indeed, it has been estimated that 45 percent of the total intestinal gases generated in humans is made up of hydrogen. Not surprisingly, there are also several examples in the scientific literature where this combustive mixture of gases has been ignited during surgical operations. Indeed, in one case, an explosion even proved to be fatal to the unfortunate patient during a colonic polypectomy. Hans Schlegel, a professor at the Institute for Microbiology of Gottingen University in Germany, explained the relationship of hydrogen as a source of energy for biological organisms:

> *"All aerobic (i.e., requiring oxygen) organisms derive the energy necessary for the construction of their cell substance and to maintain their life functions from the reaction of hydrogen and oxygen. . . . Man as well derives his metabolic energy through the slow combustion of hydrogen, although he is not being offered hydrogen in its gaseous state as nourishment, but rather as part of his foodstuffs [in which] it is weakly bonded to carbon."*

Table 5.1 provides a list of some of the most abundant elements found both in the universe and the human body. Note that the four elements that make up amino acids (i.e., hydrogen, carbon, nitrogen and oxygen) account for 99.9 percent of the elements in the universe and 98.7 percent of the elements that make up the human body.

Table 5.1: Relative Abundance of Elements

Element	Universe	Human Body
Hydrogen	90.79%	60.30%
Carbon	9.08	10.50
Nitrogen	0.04	2.42
Oxygen	0.05	25.50
Subtotal	**99.96%**	**98.72%**
Sodium	0.00012%	0.73%
Magnesium	0.00230	.01
Aluminum	0.00023	----------
Silicon	0.026	0.00091
Phosphorus	0.00034	0.134
Sulfur	0.000910	.132
Chlorine	0.000440	.032
Potassium	0.000018	0.036
Calcium	0.00017	0.226
Iron	0.0047	0.00059

From a biological perspective, each individual person is made up of some 100 trillion cells, each of which contains tens of thousands of nanobial organisms that still continue to use the basic process of hydrogen shuffling that was initially developed by their ancestors some 3.5 billion years ago. In any case, it should be clear that from both an astronomical and a biological perspective, the primordial relationships that exist between solar energy and hydrogen are symbiotic. In much the same way, if one seriously proposes using renewable energy technologies to replace the use of fossil and nuclear fuels, the relationship between solar energy and hydrogen appears to be equally inseparable, because one cannot effectively work without the other. This is because the hydrogen needs to have some sort of a primary energy input (either solar or conventional fossil or nuclear fuels) in order to break the chemical bonds, thereby allowing the hydrogen to be separated from the other atoms of oxygen or carbon. Solar energy, on the other hand, will not be able to replace fossil or nuclear-fueled energy systems unless it can be efficiently stored, transported and used as a combustion fuel in vehicles and power plants.

The Water-Former

Hydrogen is the simplest, lightest and most abundant of the 92 regenerative elements in the universe. As an energy medium, it can never be exhausted because it recycles in a relatively short amount of time. Hydrogen was first discovered in 1766 when the English chemist Henry Cavendish observed what he referred to as "inflammable air" rising from a zinc-sulfuric acid mixture. It was identified and named by the eighteenth century father of chemistry, Antoine Lavoisier, who demonstrated that Cavendish's inflammable air did indeed burn in air to form water. He concluded it was a true element, and called it hydrogen, which is a Greek word that means "water former." According to Abraham Lavi and Clarence Zener, two distinguished engineering professors at Carnegie-Mellon University, located in Pittsburgh, Pennsylvania, if hydrogen were used for electric power generation instead of the existing fossil fuels, overall electricity costs could be reduced by as much as fifty percent. The reasons that Lavi and Zener cite for this reduction are as follows:

1. Hydrogen can be burned in a combustion chamber rather than a more conventional boiler, thus high-pressure superheated steam can be generated and fed directly into a turbine. This technique reduces the capital cost of a power plant by 50 percent.

2. When hydrogen is burned, virtually no chemical pollution is generated. Thus, expensive pollution-control equipment, which can amount to one-third of the capital costs of conventional fossil fuel power plants, is unnecessary.

3. The utilization of hydrogen fuel would allow power plants to be located in the vicinity of residential and commercial loads, thereby eliminating most of the power transmission costs and line losses. This would also allow waste heat to be efficiently utilized by residential and commercial buildings.

The fact that hydrogen burns cleanly and reacts completely with oxygen to produce water makes it a more desirable fuel than fossil fuels for virtually all industrial processes. One example would be the direct reduction of iron or copper ores by hydrogen rather than by coal that would be used in a blast furnace. Another example is that hydrogen could be used not only with conventional vented burners, but with unvented burners as well. This is important because nearly 30 to 40 percent of the combustion energy of conventional burners is vented as heat and combustion by-products.

The Universal Fuel

Hydrogen is not just another energy option like oil, coal, nuclear or solar. It is unique and stands alone because it is an inexhaustible "universal fuel" that can unite virtually all energy sources with all energy uses. Although solar technologies and resources have many attractive aspects, such as being renewable, modularized, and generally pollution-free, they also tend to have equally significant disadvantages, such as not being present in the right intensity at the right place at the right time.

This lack of dispatchability is one of the major reasons why solar technologies have not been economically competitive with nuclear or fossil-fueled facilities. The hydrogen variable, on the other hand, fundamentally alters the global energy equation because it provides a realistic method of storing, transporting, and using the massive, but intermittent, supply of solar energy. As Figure 5.8 below indicates, hydrogen is a remarkable substance that is already used extensively in a wide-range of energy, industrial and chemical operations.

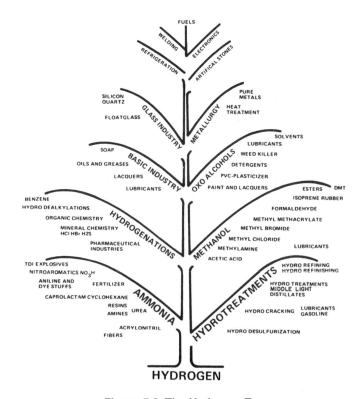

Figure 5.8: The Hydrogen Tree

Hydrogen is a primary chemical feedstock in the production of combustion fuels (including gasoline), as well as lubricants, fertilizers, plastics, paints, detergents, electronics and pharmaceutical products. It is also a premium metallurgical refining agent and an important food preservative. The image was provided courtesy of Air Products & Chemicals.

Primary vs. Secondary Energy Sources

Hydrogen can be extracted from a wide range of sources because it is in almost everything, from biological tissue and DNA, to petroleum, gasoline, paper, human waste, or water. It can be produced using electricity generated from nuclear plants, solar plants, wind plants, ocean thermal power plants or green plants. In addition, hydrogen and electricity are directly complementary, in that one can be converted into the other. Thus hydrogen is a kind of energy currency that does not vary in quality with respect to origin or location. As a result, a molecule of hydrogen made by the electrolysis of water is identical to hydrogen manufactured from other processes. These are some of the reasons why the term "hydrogen economy" is often used.

Hydrogen is often referred to as a secondary energy carrier, rather than a primary energy source, because energy must be initially used to extract the hydrogen from water, natural gas, or other compounds that contain hydrogen. This classification is misleading because it assumes solar, coal, oil or uranium to be "primary" energy sources, implying that energy is not needed to extract them from the natural environment prior to their use. Nuclear plants, for example, do not exist naturally. Indeed, they require vast energy expenditures and billions of dollars of investment before they can produce a single kilowatt of power. The same is true for the other so-called primary energy sources, including coal, oil and natural gas. Although coal and natural gas come closest to being true primary energy sources that can be burned directly with little or no refining, energy must still be expended to extract these "natural" energy sources and deliver them to the place the energy is needed. Even if drilling for oil were not required, energy must still be used to refine it into a useable energy form such as gasoline or diesel fuel. Buckminster Fuller calculated the true cost of a barrel of oil in an article, "Geoview," published in *World Magazine* in 1973:

"Scientific calculation shows that the amount of time and energy invested by nature to produce one gallon of petroleum 'safely deposited' in subterranean oil wells, when calculated in foot-pounds of work and chemical time converted to kilowatt hours and the present commercial rates at which electricity is sold amounts to approximately one million dollars per gallon."

Moreover, current energy cost accounting methods do not take into account the many global environmental problems that are a direct result of finding, transporting and burning the fossil fuels. In sharp contrast, when hydrogen is used as a fuel, its combustion by-product is essentially water vapor. If hydrogen is burned in air, which is mostly nitrogen, oxides of nitrogen can be formed as they are in gasoline and other hydrocarbon-fueled engines. But these undesirable by-products can to a large extent be eliminated in hydrogen-fueled engines by lowering the combustion temperatures of the engine (for example, by injecting water into the cylinder). Consequently, hydrogen is the cleanest-burning combustion fuel that could be used for transportation and other industrial applications.

Laboratory tests by BMW and others have shown that the air coming out of a hydrogen-fueled engine in an urban area is actually cleaner than the air that entered the engine. Accordingly, if all of the automobiles, trucks, buses, aircraft and trains were using gaseous or liquid hydrogen fuel, the air pollution in the major urban areas would be, to a large extent, eliminated. In addition, the especially serious problems of acid rain, stratospheric ozone depletion, and carbon dioxide accumulations could all be dramatically impacted by the use of hydrogen for automotive and industrial applications. Thus, an industrial transition to a hydrogen energy system may be the only realistic solution to resolving the potentially catastrophic environmental problems that threaten the Earth's biological life support systems.

Hydrogen Storage Systems

Once hydrogen has been produced, it must then be stored as either a compressed gas, a cryogenic liquid, or in a solid hydride material. For large-scale utility applications, the preferred method may be to store gaseous hydrogen underground in depleted aquifers, or depleted petroleum or natural gas reservoirs, as well as man-made caverns that have resulted from mining operations. In the case of automotive applications, it is possible to store hydrogen as a high pressure gas, although the storage tanks are bulky and the vehicle's range is typically limited to about 50 to 70 miles, which is roughly the same range as a battery powered electric vehicle.

Hydrogen Hydrides

Hydride materials absorb hydrogen like a sponge, and then release it, usually when heat is applied. While there are hundreds of potential hydride material selections, the most common hydride systems that have thus far been used in automotive vehicles have consisted primarily of metal particles of iron and titanium that were initially developed by researchers at Brookhaven National Laboratory. In tests conducted by American investigators at Billings Energy Corporation (Provo, Utah) and German investigators at Daimler-Benz (Stuttgart, Germany), these hydride systems have been shown to be a very safe method of storing hydrogen in automobiles, but they are about 18 to 20 times heavier than liquid hydrogen storage systems.

Figure 5.9: Iron-titanium hydrogen hydrides
Hydrogen hydride alloys are milled into the small gray particles that are pictured above in the cutaway of the hand-held hydride storage tank. The image was provided courtesy of the Tappan Appliance Company.

Figure 5.10: Hydrogen-Fueled Mercedes Benz
Helmut Buchner, Daimler-Benz hydrogen-project leader, displays a
model of a hydride storage system used in the vehicle behind him.

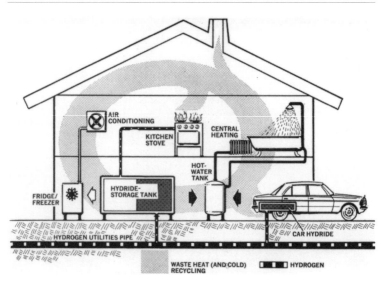

Figure 5.11: Hydrogen Residential Delivery System
Both photographs are reprinted from Popular Science with permission
from Times Mirror Magazines, Inc.

There are other hydrogen-storage materials under development that do not have such severe weight penalties as the current generation of metal hydrides, such as lightweight, high surface area carbon composites that have the ability of storing gaseous hydrogen with weight and volume characteristics that are comparable to existing gasoline tank systems. However, with the present iron-titanium hydride systems, if a typical range of 500 kilometers (310 miles) were to be provided, the storage system would weigh about 2,600 kilograms (5,700 pounds). In contrast, a liquid hydrogen tank providing a similar range would weigh about 136 kilograms (300 pounds).

A comparable gasoline tank would weigh about 63 kilograms (138 pounds). If an electric vehicle with conventional lead-acid batteries were to have a 500-kilometer (310 mile) range, the weight of the batteries would be in excess of 2,000 kilograms (4,400 pounds). More efficient battery systems are being developed by numerous manufacturers, but the most efficient electric vehicles in the future may not be energized by batteries at all -- but by "fuel cell" systems that electrochemically convert hydrogen and oxygen directly into electricity, with pure water as a by-product. Fuel cells could make electric vehicles truly practical by eliminating the heavy battery storage systems that are presently required in electric vehicles. Fuel cells were first developed in England in 1839 by Sir William Grove, but they were not utilized for practical applications until the 1960s when NASA utilized them in manned spacecraft to provide electricity as well as the drinking water for the astronauts.

A number of major automotive companies are now actively developing fuel cell powered vehicles, and all fuel cells need a particular type of fuel -- hydrogen. Although one might assume that before fuel cell-powered vehicles become mass-produced, hydrogen fuel would have to become readily available at refueling stations, this assumption is incorrect. This is because it is possible to extract the hydrogen from hydrocarbon fuels, such as gasoline. Other manufacturers, such as BMW and Ford, are not convinced that fuel cell powered vehicles will be able to displace internal combustion engines that are optimized to operate on both hydrogen as well as gasoline. Moreover, the extensive safety and crash tests involving liquid hydrogen storage tanks and related components have shown that liquid hydrogen is a much safer fuel to use than gasoline.

Figure 5.12: Hydrogen-Fueled BMW Engine
BMW has modified conventional internal combustion engines to operate on both hydrogen as well as gasoline. These "bivalent" engines include the V-12 engine pictured below.

ELECTROLYZER
Combines Electricity
& Water

to Make Gaseous
Hydrogen
and Oxygen
($H_2 + O_2$)

POLLUTION-FREE FUEL
from
SUNSHINE & WATER

DISH STIRLING GENSET
Generates Solar Electricity

STIRLING CRYOCOOLER
Produces Liquid Hydrogen (LH_2)

BMW WITH LH_2 FUELING STATION

Figure 5.13: Hydrogen Renewable Fuel Cycle

Liquid Hydrogen

In order for hydrogen gas to be liquefied, it needs to be cooled to minus 421.6 degrees Fahrenheit. This means that if a beaker of liquid hydrogen were sitting on a table at room temperature, it would be boiling as if it were water sitting on a hot stove. If the beaker of liquid hydrogen were spilled on the floor, it would be vaporized and dissipated in a matter of seconds. If liquid hydrogen were poured on one's hand, it would only feel slightly cool to the touch as it would slide through one's fingers and fall to the ground. This is due to a thermal barrier that is provided by the skin. On the other hand, if one placed one's hand in a vessel containing liquid hydrogen, severe injury would occur in seconds because of the ultra-cold temperatures. However, such a circumstance would be unlikely in typical automotive environments where the main concern involves refueling the vehicle and collisions. In most automotive accidents, the most serious concern is the possibility of a fuel-fed fire and/or explosion, and under such circumstances, gaseous or liquid hydrogen would in almost all cases be the preferred fuel if one were seeking to minimize injury to the occupants of the vehicle.

Liquid hydrogen has the lowest weight per unit of energy of any fuel; it has relatively simple supply logistics; and it has refuel time requirements comparable to gasoline fuel systems. Unlike the current hydride energy storage options, liquid hydrogen is similar to gasoline in terms of space and weight on-board a vehicle. Although a liquid hydrogen storage tank for an automobile is expected to be approximately twice as heavy than a typical 30-pound gasoline tank, in larger vehicles that carry greater volumes of fuel, such as trucks, trains or aircraft, the difference in tank weight will be more than offset by the difference in fuel weight. For example, studies by Lockheed have shown that a large commercial aircraft would have its overall takeoff weight reduced by as much as forty percent if liquid hydrogen were used instead of conventional aviation kerosene. Although liquid hydrogen fuel offers many safety advantages in vehicular applications, it also has significant technical issues that need to be put into perspective. Low-temperature cryogenic fuels like liquid hydrogen are more difficult to handle and substantially more difficult to store, in contrast with conventional hydrocarbon fuels like gasoline or aviation kerosene that are liquid at ambient temperatures.

Figure 5.14: A Liquid Hydrogen-Fueled Buick
Walter Stewart and his colleagues at Los Alamos National Laboratory
modified this 1979 Buick to operate on liquid hydrogen fuel in 1982.

Figure 5.15: A Liquid Hydrogen Self-Service Refueling Station
The photographs were provided courtesy of the German Aerospace
Research Establishment and Los Alamos National Laboratory.

Figure 5.16: A Liquid Hydrogen-Fueled BMW
BMW is the first major automobile manufacturer to modify a vehicle to utilize a liquid hydrogen fuel system. The photograph courtesy of BMW.

Figure 5.17: An Automated Hydrogen Refueling Station
The world's first hydrogen filling station in Munich, Germany features a robotic attendant that allows the driver to remain seated during the refueling operation that takes approximately four minutes.

Figure 5.18: A Liquid Hydrogen Storage Tank
The photograph was provided courtesy of BMW.

Figure 5.19: Four Generations of Liquid Hydrogen-Fueled BMWs

With the initial double-walled, vacuum-jacketed storage tank that was used in a 1979 Buick by investigators at Los Alamos National Laboratory, the liquid hydrogen would evaporate at a rate of about 10 percent per day if the car was not driven. This means a full tank of liquid hydrogen could evaporate in about 10 days. Recent advances in Germany, however, have reduced evaporation rates of cryogenic tanks to less than two percent per day, and investigators in Japan claim to have reduced the losses to less than one percent per day. Venting of the fuel needs to occur to keep the fuel tank from rupturing due to a pressure build up, and it not only presents the problem of an empty fuel tank, but if the vehicle is in an enclosed space such as a garage, the vented hydrogen could pose a risk of being ignited due to hydrogen's wide flammability limits. Although hydrogen explosions are rare, any combustible gas in an enclosed space, such as an underground parking structure is an obvious safety concern.

As an interim solution to this problem, German engineers at BMW initially installed a small burner underneath their liquid hydrogen-fueled BMW test vehicles. The small burner is continuously ready for operation and, upon reaction of the pressure release valve, a pilot flame will safely combust the escaping hydrogen. However, in the fifth-generation 750 hL "hydrogen 7 series" BMWs, which are the first production hydrogen-fueled cars in the world, the relatively small quantity of hydrogen evaporating from hydrogen storage tanks is directed to a small (5 kW) fuel cell that uses the hydrogen to generate electricity. The electricity can then be used to operate the lights, recharge the battery, or provide heating or air conditioning for the vehicle's occupants while the engine is not running. A fleet of 15 of the BMW 750hL models is already on the road in Germany performing shuttle service in both Hanover and Munich.

Some of the other concerns about liquid hydrogen fuel systems for vehicles include the following:

1. Vacuum-jacketed storage tanks and the related piping and control subsystems required for liquid hydrogen are more expensive than conventional fuel tanks. The cost of a gasoline tank for an automobile is about $150, whereas a liquid hydrogen storage tank would be expected to cost from $1,000 to $2,000, assuming production volumes of 100,000 units per year.

2. Liquid hydrogen is similar to gasoline in terms of weight, but because of the energy density of liquid hydrogen, it requires a fuel tank to be roughly three to four times as large in volume as its gasoline or aviation fuel counterpart.

3. Liquid hydrogen fuel systems would require changes in the energy infrastructure, such as pipelines and fuel transportation vehicles and end-use systems, such as aircraft, stoves, automotive engines as well as fuel pumping systems.

Although a liquid hydrogen fuel tank may cost more than a gasoline tank in comparable levels of production, anyone who has seen what happens to people who have had the unfortunate experience of being burned from a gasoline fire would never argue about the additional costs of a liquid hydrogen fuel tank. While it is true that cryogenic fuels are more difficult to handle than conventional hydrocarbon fuels, a self-service liquid hydrogen pumping station that had been developed by a German consortium that included BMW and Linde, was provided to investigators at Los Alamos National Laboratory. After using this system for over a year, the Los Alamos investigators, headed up by Walter Stewart, concluded the following:

> "...liquid hydrogen storage and refueling of a vehicle can be accomplished over an extended period of time without any major difficulty."

The engineers at BMW, which have been developing liquid hydrogen-fueled cars for more than 20 years, are convinced that liquid hydrogen offers the most practical alternative to existing hydrocarbon fuels, and while the technical issues of using liquid hydrogen are substantial, they are also manageable. With the current hydrogen-fueled BMWs, the insulation in the cryogenic storage tank allows loss-free storage for about three days, and the vehicle has a range of approximately 250 miles. Moreover, if liquid hydrogen were used on all new automobiles in the U.S., production volumes of cryogenic storage tanks would not be in the thousands, but in the millions. As such, the cost of the individual cryogenic tanks would be substantially reduced.

Although a liquid hydrogen fuel tank will be three or four times larger than a conventional fuel tank, it will require only minimal design changes for most vehicles. Even in the case of aircraft, the changes are not significant. In order to minimize the amount of drag that would result from having larger wing tanks, Lockheed engineers have proposed extending the length of the fuselage of the aircraft to accommodate the larger liquid hydrogen tanks as shown in Figure 5.28. Other configurations under study by Boeing are shown in Figures 5.30 and 5.31. With respect to the increased costs associated with making a changeover to hydrogen energy systems, it needs to be remembered that the environmental costs of finding, transporting and burning fossil fuels are not calculated in the current energy pricing structure. This is in spite of the fact that the increasing concentrations of atmospheric pollution are costing billions of dollars in additional health care costs, crop losses, and the corrosion of buildings and other structures.

The Hydrogen Engine

Major automotive firms have been actively involved in hydrogen research and development to some extent for many years. Hydrogen-fueled engines have been shown to be more energy efficient because of their complete combustion. Moreover, gasoline and diesel-fueled engines form carbon deposits and acids that erode the interior surfaces of the engine and contaminate the engine oil. This, in turn, increases wear and corrosion of the bearing surfaces. Since hydrogen-fueled engines produce no carbon deposits or acids, it is anticipated they will require considerably less maintenance and have a longer operating life. In addition, hydrogen fuel can also be utilized with more efficient Stirling cycle engines that could allow hybrid fuel cell electric vehicles to be truly practical. Automotive engineers have been aware of hydrogen's favorable combustion characteristics since the early 1900s. A German engineer, Rudolf A. Erren, began optimizing internal combustion engines to use hydrogen in the early 1920s, and dozens of his modified vehicles were used in Berlin. As a result, Erren is generally recognized by the hydrogen technical community as being the father of the hydrogen-fueled engine.

Figure 5.20: Rudolf A. Erren
Rudolf Erren, a German engineer, is generally considered to be the father of the hydrogen engine. The photo was taken in Germany in 1979 and is reprinted with permission from Peter Hoffmann, author of *The Forever Fuel, The Story of Hydrogen.*

Erren modified a wide range of trucks and buses, and Allied forces even captured a German submarine in World War II that had not only a hydrogen-fueled engine, but hydrogen-powered torpedoes as well that were initially designed and patented by Erren and his associates.

Roger Billings

The first hydrogen-fueled automobile in the U.S. was a Model A Ford truck, developed in 1966 by Roger Billings, a rather young, but quite remarkable investigator. At the time, Billings was still a student in high school, but several years later as an undergraduate student in chemistry at Brigham Young University, he won a 1972 Urban Vehicle Design Contest with a hydrogen-fueled Volkswagen. Billings eventually established Billings Energy Corporation in Provo, Utah and went on to modify a wide range of automotive vehicles, including a Winnebago motor home that not only had the engine fueled by hydrogen, but the electrical generator and all of the vehicle's appliances. Billings went on to design and construct the world's first "hydrogen homestead," pictured in Figure 5.22, which has virtually all of the home's appliances modified to operate on hydrogen.

Figure 5.21: Roger Billings
The photograph was provided courtesy of the
American Academy of Science.

A component of the Billing's hydrogen home is a dual-fueled Cadillac Seville that can operate on both hydrogen and gasoline as well as a hydrogen-fueled Jacobsen farm tractor. The driver of the dual-fueled vehicle is able to change from hydrogen fuel to gasoline while driving with the flip of a switch from inside the vehicle. As a result of his many years of research, Billings and his colleagues have been able to demonstrate that hydrogen could indeed be used as a universal fuel in end-use applications. The photograph in Figure 5.23 shows a special burner head that had been modified by the Tappan Company for hydrogen combustion. Because hydrogen burns with an invisible flame, it was necessary to utilize a material that could allow someone to know the range was on. The engineers at Tappan utilized a steel wool catalyst that rests on the burner head to solve this problem.

Figure 5.22: A Hydrogen Homestead
All of the home's appliances and vehicles have been modified to utilize hydrogen fuel. The photograph was provided courtesy of the American Academy of Science.

Figure 5.23: A Tappan Hydrogen-Fueled Gas Range
The stainless steel mesh on the hydrogen-fueled range glows when it is heated, resembling an electric range surface when the burner is on. The steel wool is covered by a stainless steel shroud (shown to the left) that slips over the catalyst and burner head to give a clean, uncluttered appearance. Photograph courtesy of the Tappan Appliance Company.

Figure 5.24: A Hydrogen-Fueled Coleman Stove
The photograph was provided courtesy of the
American Academy of Science.

To demonstrate that even small portable appliances can be modified to use hydrogen fuel, Billings Energy Corporation also adapted a small portable Coleman Stove pictured in Figure 5.24 for hydrogen fuel use. The small hydrogen storage tank on the right of the stove utilizes iron-titanium metal hydrides. Because of anticipated fossil fuel shortages, hydrogen research programs have been undertaken by the U.S. Air Force, Navy and the Army since the 1940s (refer to Figures 5.25, 5.32 and 5.35). However, because of the perceived difficulties of storing liquid hydrogen on-board automotive vehicles, and the fact that it was more expensive to produce than the hydrocarbon fuels that were being refined from oil, the hydrogen fuel option was not seriously pursued. At the time, the Department of Defense was not concerned with environmental considerations, and since it has to purchase fuel on the open market like every other consumer, fuel costs were the primary concern. Prior to the Arab oil embargo in 1973, oil was selling for less than three dollars a barrel, and as long as there was not an emergency, the fuel supply problem was essentially left to the oil companies to worry about.

Figure 5.25: A Liquid Hydrogen-Fueled Tank
Specifications for a M-60 tank fueled with diesel fuel, in contrast to liquid hydrogen.

Diesel / Length: 27..3 feet / Weight: 114,000lbs

Liquid Hydrogen / Length: 36.8 feet / Weight: 130,171 lbs

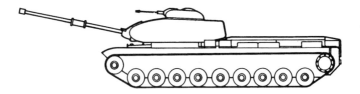

Hydrogen Safety

Many individuals believe hydrogen is especially dangerous. Some people think hydrogen energy is somehow related to the "hydrogen" bomb. This is understandable given the common usage of the word hydrogen, but when hydrogen is used as a fuel, a simple chemical reaction that involves the transfer of electrons occurs. A hydrogen bomb, on the other hand, involves a high-temperature nuclear fusion reaction similar to that which occurs in stars.

Other people remember that the great German airship, the *Hindenburg*, was using hydrogen as a lifting-gas when it was destroyed by fire in a spectacular crash in New Jersey in 1937 that was captured on film. This event had a profound emotional impact on a great many people, but while 35 people died in the unfortunate accident, it is hardly ever mentioned that 62 other people survived. Although the cause of the *Hindenburg* fire was a mystery for may years, Dr. Addison Bain, the former head of hydrogen safety programs for NASA, has been able to determine that the fire on board the was not caused by the hydrogen lifting gas, nor was the hydrogen responsible for the rapid spread of the fire. After a careful analysis of the accident data, Bain was able to determine that the *Hindenburg* fire was caused by static electricity that reacted with the highly inflammable aluminum-based paint mixture that was used to protect the outer skin of the *Hindenburg* from ultraviolet radiation. When the thunderstorm caused an electrical discharge to occur, this highly reactive paint, which had the chemical properties of rocket fuel, ignited and then was responsible for the swift spread of the flames. Thus while some of the hydrogen gas was ignited in the process, it was not the primary cause of the *Hindenburg* accident.

It is also rarely mentioned that prior to its fatal crash in 1937, the *Hindenburg* had successfully completed ten round trips between the United States and Europe, and its sister ship, the *Graf Zeppelin*, had made regular scheduled transatlantic crossings from 1928 through 1939 with no mishaps. Indeed, of the 161 rigid airships that were built and flown between1897 and 1940 (nearly all of which used hydrogen as a lifting gas), only 20 were destroyed by fires. In addition, of the 20, seventeen were lost in military incidents that in many cases resulted from hostile enemy fire during World War I. That is an excellent safety record for the technology of the day.

Hydrogen has a wider range of flammability when compared to gasoline. For example, a mixture as low as 4 percent hydrogen in air, or as high as 74 percent, will burn, whereas the fuel to air ratios for gasoline are only from 1 to 7.6 percent. In addition, it takes very little energy to ignite a hydrogen flame, about 20 micro-joules, compared to gasoline, which requires 240 micro-joules. However, the hazardous characteristics of hydrogen are offset by the fact that it is the lightest of all elements. Because the diffusion rate of a gas is inversely proportional to the square root of its specific gravity, the period of time in which hydrogen and oxygen are in a combustible mixture is much shorter than for hydrocarbon fuels. The lighter the element is, the more rapidly it will disperse if it is released in the atmosphere. In the event of a crash or accident where hydrogen is released, it would rapidly disperse up and away from people and other combustible material within the vehicle.

Figure 5.26: The Hindenburg
The Hindenburg would dwarf a contemporary Boeing 747 jetliner.

Figure 5.27: The Accident
It is generally not known that while 35 people died in the unfortunate *Hindenburg* accident, 62 other people survived the event. Moreover, according to the accident report, of the 35 people that died in the accident, 33 died because they jumped from the airship when it was more that 100 feet in the air -- and they died from the fall. While there were two people who were burned to death, they were burned by the diesel fuel that the Hindenburg was using to power its Mercedes Benz engines.

Hydrogen Explosions

The *Hindenburg* did not explode, as most people assume. Rather, it caught fire, and as the flames rapidly spread, the airship sank to the ground. Although sabotage was suspected, most experts believe the fire was started because the airship was venting some of its hydrogen (in order to get closer to the ground) in the middle of an electrical thunderstorm. In addition, the airship was simultaneously moored to the ground by a steel cable, which only increased the risk of the hydrogen being ignited by static electrical discharge. Hydrogen explosions are extremely powerful when they occur, but they are extremely rare. This is because hydrogen needs to be in a confined space for an explosion to occur. Out in the open, it is almost impossible to get hydrogen to explode without the use of a heavy blasting cap.

In 1974, Paul M. Ordin, a research analyst working for NASA presented a paper to the Ninth Intersociety Energy Conversion Engineering Conference, which reviewed 96 accidents or incidents involving hydrogen. NASA cryogenic tanker trailers had transported more than 16 million gallons of liquid hydrogen for the Apollo-Saturn program alone, and while most mishaps were of a highly specialized nature, there were five serious highway accidents that involved extensive damage to the liquid hydrogen transport vehicles. These accidents were such that if conventional gasoline or aviation kerosene had been involved, a spectacular blaze would have been expected, but due to the physical characteristics of liquid hydrogen, none of the accidents resulted in either a hydrogen explosion or fire. This is partly why NASA investigators rate hydrogen to be safer than gasoline.

The U.S. Defense Department has been conducting liquid hydrogen-fuel safety research since 1943. In tests undertaken by the Air Force Flight Dynamics Laboratory at Wright-Patterson Air Force Base, armor-piercing incendiary and fragment-simulator bullets were fired into aluminum storage tanks containing both kerosene and liquid hydrogen. The test results indicated that the liquid hydrogen was actually much safer than conventional aviation kerosene in terms of gross response to a ballistic impact. Other tests involved simulated lightning strikes, using a 6-million volt generator that shot electrical arcs directly into the liquid hydrogen fuel containers. In neither of these tests nor the earlier Defense Department tests did the liquid hydrogen explode.

Fires did ignite as a result of the simulated lightning strikes, but in the case of liquid hydrogen, the fires were less severe even though the total heat content of the hydrogen was twice that of kerosene. These are some of the main reasons why liquid hydrogen would be a much more desirable fuel than conventional fossil-based fuels in combat situations where an aircraft's fuel tanks could be penetrated by explosive bullets or fragments.

One example of a dangerous situation where explosive mixtures of hydrogen and oxygen were present in a confined space occurred when the highly publicized partial meltdown occurred in 1979 at the Three Mile Island (TMI) nuclear facility in Pennsylvania. Nuclear reactors operate at very high temperatures, and the only thing that prevents their six to eight inch thick steel reactor vessels from melting is the large amounts of cooling water that must be continuously circulated in and around the reactor vessel. An average commercial-sized reactor requires approximately 350,000 gallons of water per minute. During the process of nuclear fission, the center of the uranium fuel pellets in the fuel rods heat up to about 4,000 degrees Fahrenheit. The cooling water inside a nuclear reactor is critical because it keeps the surface temperature of the nuclear fuel pellets down to about 600 degrees. If, for any reason, the water in the reactor is not present within 30 seconds the temperatures in the reactor vessel will soar to over 5,000 degrees. Such searing temperatures are not only high enough to melt steel, they will also thermochemically split any water present into a highly explosive mixture of hydrogen and oxygen. This is what happened at TMI. If a spark had ignited the hydrogen and oxygen gas bubbles that drifted to the top of the containment building at TMI, the resulting powerful explosion would likely have fractured the containment building. This, in turn, would have resulted in the release of large concentrations of radiation at ground level.

Fortunately, the hydrogen gas bubble at TMI was successfully vented without a major mishap, but as long as any of the hydrogen and oxygen remained in the confined space of the containment building, its potential for detonation was of paramount concern. It should be noted that a hydrogen gas bubble developing from a serious nuclear reactor accident is a highly unusual event, even for the current generation of nuclear-fission reactors. It is, however, an excellent example of the relatively specialized conditions that are required for hydrogen to actually explode.

Hydrogen Aircraft Applications

On March 27, 1977, two fully loaded Boeing 747 commercial aircraft crashed into each other on a foggy runway in the Canary Islands. This disaster, the worst in aviation history, took 583 lives, compared to the 35 lost in the crash of the *Hindenburg*. Investigators concluded many of the deaths in the Canary Islands accident were a result of the kerosene-fueled fire that raged for more than 10 hours. Unlike hydrogen, hydrocarbon fuels are heavy because the hydrogen is bonded to carbon, thus when hydrocarbon fuels vaporize, their gases sink rather than rise in the atmosphere. This means burning fuel falls and burns them alive. That is why hydrogen would be a more desirable vehicular fuel if a serious accident were to occur -- especially on board a commercial aircraft that has large volumes of fuel on board.

G. Daniel Brewer, who was at the time the hydrogen program manager for Lockheed, told an audience of experts six weeks after the Canary Islands accident that if both aircraft had been using liquid hydrogen fuel instead of kerosene, hundreds of lives could have been saved because of the following reasons:

1. Liquid hydrogen cannot react with oxygen and burn until it first vaporizes into a gas. As it does evaporate, it dissipates rapidly as it is released in open air. As a result, the fuel-fed portion of the fire would have only lasted a few minutes; not many hours as with conventional liquid hydrocarbon fuels.

2. If liquid hydrogen were used as fuel, the resulting fire would have been confined to a relatively small area because the liquid hydrogen would rapidly vaporize and disperse in the air, burning upward, rather than spreading laterally like kerosene.

3. Heat radiated from the hydrogen fire would be significantly less than that generated by a hydrocarbon fire. As a result, only persons and structures immediately adjacent to the flames would be affected. The hydrogen fire would produce no smoke or toxic fumes, which can be even more dangerous to the passengers and crew than the actual flames themselves.

4. With liquid hydrogen fuel storage tanks, the gaseous hydrogen that vaporizes will fill the empty volume inside the tanks. The hydrogen is not combustible because no oxygen is present. With conventional hydrocarbon fuel tanks, air fills the empty volume of the tanks and combines with vapors from the fuel to create a potentially combustible mixture that is a formidable safety hazard.

According to industry data, 80 percent of all people who die in commercial aircraft accidents do not die from the crash, but from the fire and resulting toxic fumes that the fire generates.

Figure 5.29 provides an artist's conception of a typical airline terminal at a liquid hydrogen-fueled airport. A liquid hydrogen-fueled Lockheed L-1011 commercial aircraft is in the process of being refueled by a ground crew. In the distance, there are two vacuum-jacketed liquid hydrogen storage spheres (about 35 feet in diameter). Vacuum-jacketed cryogenic fuel lines contained in a trench covered with an open steel grate carry the liquid hydrogen from the storage vessels to a series of hydrant pits, like the one being used to refuel the aircraft. A "cherry picker" truck makes the connection between the hydrant pit and the liquid hydrogen-fueled aircraft. Of the two gas lines shown, one taps off gaseous hydrogen displaced from the aircraft tanks by the incoming liquid hydrogen and returns it to the liquefaction plant that is located off in the distance.

Detailed studies by Lockheed investigators have concluded that in addition to hydrogen's favorable safety characteristics, liquid hydrogen-fueled aircraft will have other significant advantages over their fossil-fueled counterparts; they will be lighter; quieter; will require smaller wing areas; shorter runways; and will minimize pollution. They will also use less energy for two reasons: less hydrogen fuel is needed per flight mile; and less energy is required to manufacture the hydrogen fuel compared with alternative fossil fuels. Because liquid hydrogen has the highest energy content per weight of fuel, the range of an aircraft could be roughly doubled, even though its takeoff weight would remain essentially the same. Because of the lack of carbon in hydrogen-fueled aircraft, the operation and maintenance costs on the engines will be substantially reduced, and the noise during takeoff may also be reduced by roughly one-third.

Figure 5.28: A Liquid Hydrogen-Fueled Lockheed Aircraft Design
The overall dimensions and characteristics of a liquid hydrogen-fueled aircraft proposed by Lockheed Aircraft Corporation. The Illustration was provided courtesy of Lockheed Missiles & Space Company.

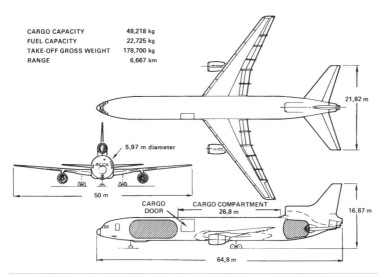

CARGO CAPACITY	48,218 kg
FUEL CAPACITY	22,725 kg
TAKE-OFF GROSS WEIGHT	178,700 kg
RANGE	6,667 km

5,97 m diameter

21,82 m

50 m

CARGO DOOR   CARGO COMPARTMENT 26,8 m   16,87 m

64,8 m

Figure 5.29: A Liquid Hydrogen-Fueled Airport
An artist's illustration of a liquid hydrogen-fueled commercial airport. The illustration provided courtesy of Lockheed Missiles & Space Co.

Figure 5.30: A Liquid Hydrogen-Fueled Boeing 747
A Boeing 747 modified to carry all liquid hydrogen fuel in the expanded
upper lobe of the aircraft's fuselage.

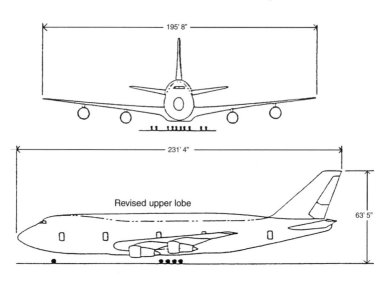

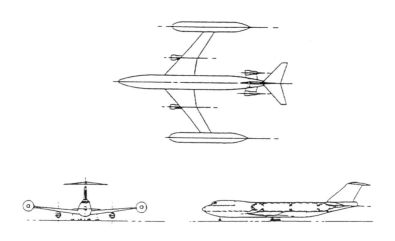

Figure 5.31: Liquid Hydrogen Wing Tanks
A Boeing 747 configured with hydrogen fuel tanks in wing tip nacelles.

Figure 5.32: A Liquid Hydrogen-Fueled Military Transport Aircraft
The U.S. Air Force has been evaluating liquid hydrogen-fueled aircraft since the 1940s. The images below provide a comparative fuselage cross-section of conventional aviation kerosene and liquid hydrogen-fueled military transport aircraft.

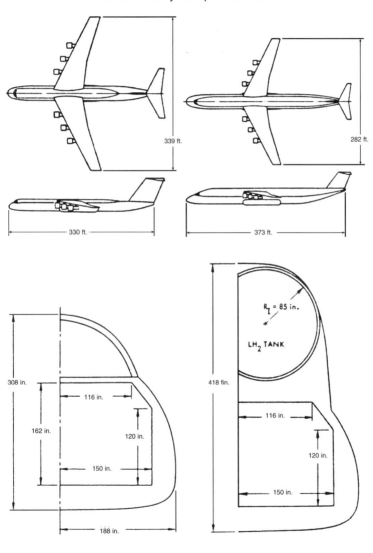

Aviation Kerosene
(Wt.: 1,839,000 lbs.)

Liquid Hydrogen
(Wt.: 1,275,000 lbs.)

Figure 5.33: Liquid Hydrogen-Fueled Freighter
A Lockheed designed liquid hydrogen-fueled freighter will be able to carry up to two million pounds of cargo over a range of 16,000 miles without refueling. The image provided courtesy of *Popular Mechanics*.

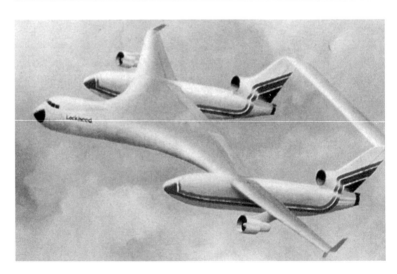

Figure 5.34: Lockheed Liquid Hydrogen-Fueled Freighter
Lockheed engineers developed multiple concepts for future liquid hydrogen-fueled freighters, like the ones pictured above. The image was provided courtesy of Popular Mechanics.

Figure 5.35: Air Force Liquid Hydrogen-Fueled B-57
Converted by NASA in 1956, this was the first aircraft in the United
States to operate with liquid hydrogen fuel.

Figure 5.36: Liquid Hydrogen-Fueled Space Shuttle
The main engines on the space shuttle are fueled with liquid hydrogen.
The solid fuel boosters on the side of the shuttle use an aluminum-
based fuel that results in toxic emissions that are clearly visible in the
picture. The photograph was provided courtesy of NASA.

NASA

Werner Von Braun, a German rocket engineer who helped to develop the V-2 rockets in World War II, directed the first engineering effort to utilize liquid hydrogen as a rocket fuel. After the war, Von Braun was responsible for helping to direct the U.S. space program, which eventually evolved into the National Aeronautics and Space Administration (NASA). Because liquid hydrogen has the greatest energy content per unit weight of any fuel, NASA engineers selected liquid hydrogen as the primary fuel for the Saturn 5 moon rockets and the Space Shuttle. NASA also funded research by numerous aerospace firms, including Lockheed and Boeing, to determine if liquid hydrogen could be economically used in conventional commercial aircraft, and what modifications would need to be made to airports and related fuel systems if liquid hydrogen were to be used as a commercial aircraft fuel.

Figure 5.37: A Liquid Hydrogen-Fueled Launch Vehicle
NASA engineers initially planned to utilize the liquid hydrogen-fueled manned shuttle launch vehicle pictured above, but Congressional budget cuts forced the use of solid-rocket boosters instead.
The image was provided courtesy of NASA.

Figure 5.38: Separation
Separation of space shuttle from manned launch vehicle. The illustration was provided courtesy of NASA.

Figure 5.39: Return of Manned Launch Vehicle
An artist's concept of the shuttle's manned liquid hydrogen-fueled launch vehicle returning to Earth for a landing. The illustration was provided courtesy of NASA.

Figure 5.40: The Hydrogen-Fueled Saturn 5 Moon Rocket
The photograph was provided courtesy of NASA.

Hypersonic Aerospacecraft

While liquid hydrogen fuel offers many advantages for sub-sonic aircraft, its advantages increase substantially for superson-ic, and the much more advanced aircraft that have been referred to as "hypersonic aerospacecraft." Such advanced vehicles have been under development in the U.S., Germany and Great Britain. In the U.S., a number of governmental agencies, including NASA, the Navy and the Air Force have all been actively researching and developing various subsystems for such advanced aircraft for many years. When they become operational, these advanced transport vehicles will make the current generation of space shut-tles essentially obsolete. Hypersonic aircraft will be able to take off from conventional runways, achieve speeds in excess of Mach 15, ascend into Earth orbit and fly back to the Earth in powered flight. While such capabilities would have important military advantages, such spacecraft would also accelerate the commer-cialization of space.

Whereas the Space Shuttle relies on expensive ceramic tiles to keep the surface of the spacecraft from melting during reentry, as Figure 5.45 indicates, the hypersonic aerospacecraft will use the ultra-cold liquid hydrogen fuel to cool its outer surface. The hypersonic aircraft pictured in Figure 5.41 was initially designed by Lockheed in the 1970s as an advanced supersonic transport (SST). Such a vehicle was being designed to carry about 200 passengers at speeds in excess of Mach 6 (i.e., about 4,000 miles per hour). Figure 5.42 provides a more updated version of an aerospacecraft, referred to as the X-30 (experimental) National Aerospace Plane (NASP). If such an aircraft were ever built, it could travel from New York to Los Angeles in about 12 minutes at speeds in excess of 8,000 miles per hour. Although commercial interest in the SST program did not materialize in the U.S. in the 1960s, the U.S. Air Force continued funding advanced SST and hypersonic research because of the obvious military implications of such technology. As a result of this research, aerospace engineers in both the U.S. and Great Britain have made enough technical progress in supercomputers and propul-sion systems to actually fabricate test systems. However, the development of such technologies is an extremely complex undertaking, and many billions of dollars will have to be expend-ed before such vehicles will ever be commercially viable.

Figure 5.41: Hypersonic Aircraft
An artist's concept of a 4,000 mph liquid hydrogen-fueled hypersonic aircraft. The image was provided courtesy of Lockheed Missiles & Space Corporation.

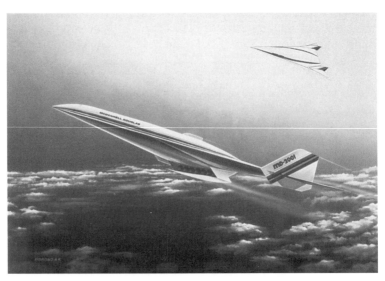

Figure 5.42: McDonnell Douglas Aerospace Plane
The McDonnell Douglas 8,000 mph X-30 liquid hydrogen-fueled National AeroSpace Plane. The image was provided courtesy of McDonnell Douglas Corporation.

Figure 5.43: Hypersonic Aircraft Cutaway
A cutaway of the McDonnell Douglas X-30 hypersonic aircraft shows
the liquid hydrogen fuel storage and integrated surface cooling system.
The image was provided courtesy of McDonnell Douglas Corporation.

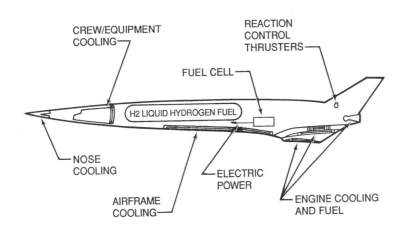

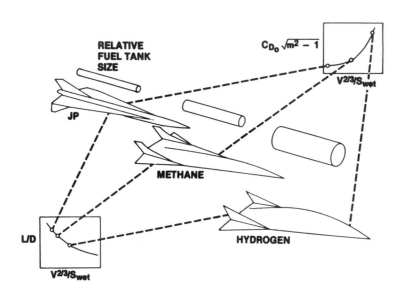

Figure 5.44: Relative Fuel Densities
Fuel selection impacts size, weight & drag. The image was provided
courtesy of McDonnell Douglas Corporation.

Figure 5.45: Liquid Hydrogen Cooling System
Hypersonic aerospacecraft will use the ultra-cold liquid hydrogen fuel to cool its outer surface. The image was provided courtesy of McDonnell Douglas Corporation.

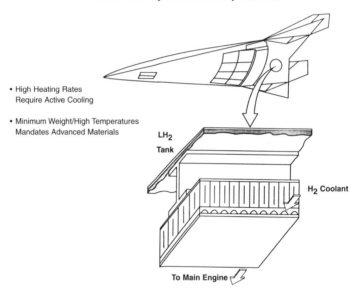

Space Habitats

While hypersonic vehicles offer a glimpse of what advanced hydrogen technologies will look like, the ultimate hydrogen ener-gy transportation -- and living -- technology will be hydrogen-fueled space habitats. Even with a successful energy transition to renewable resources, it is only reasonable to assume that the other available resources on the Earth are limited; and as the human population continues to exponentially increase, it will become increasingly necessary to utilize the vast resources that exist in space. Dr. Gerard K. O'Neill, a professor of physics at Princeton University, has pointed out that the resources that abound in space are virtually inexhaustible. In his paper, "The Colonization of Space" published in *Physics Today* and his relat-ed book, *The High Frontier: Human Colonies in Space,* he calcu-lated that it would be possible to have more humans living in large space colonies than on the Earth by the year 2075.

O'Neill and his colleagues did detailed studies on the construction of large-scale space habitats that would rotate to simulate Earth's gravity. This means large cities in space would allow humans and other mammals to remain in space for extended periods of time without having adverse physiological reactions. Without simulating the gravity of Earth, for example, the human body would soon turn into a shapeless, jellyfish type of organism. One of the most far-sighted hydrogen-fueled spacecraft configurations has been conceived by R.W. Bussard, a member of the British Interplanetary Society. The Bussard spacecraft pictured in Figure 5.46 is called an interstellar ramjet "hydrogen scoop." This is because it is designed to scoop up free hydrogen atoms that drift in space between the stars and then accelerate them into fusion engines for propulsion. Such a configuration represents a renewable-fueled hydrogen spacecraft that could accelerate to nearly the speed of light.

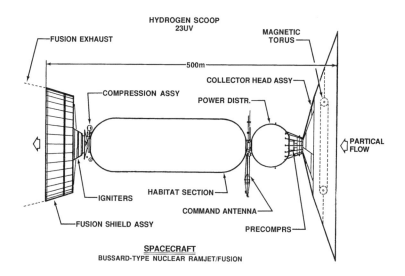

Figure 5.46: Hydrogen Scoop
Blueprints for a Bussard "Hydrogen-Scoop" Ramjet Spacecraft.
The illustration was prepared by Rick Sternback.

It has been estimated by Carl Sagan in his book *Cosmos,* that in deep space, there is only about one hydrogen atom per ten cubic centimeters, which is a volume about the size of a grape. This means that if a hydrogen-fueled Bussard ramjet is to be able to work, it would need a frontal scoop that is perhaps 500 kilometers (i.e., 310 miles) across, and the engines would have to be the size of small cities. If O'Neill's assumptions and calculations are correct, building large structures in space may be a straightforward engineering task. However, building the fusion-powered engines capable of accelerating a spacecraft to nearly the speed of light is another matter. Even if the daunting technical obstacles of achieving such speeds were resolved, other serious problems would be created by traveling at such high speeds.

As Sagan has pointed out, hydrogen and other atoms would then be impacting the spacecraft at close to the speed of light, and such induced cosmic radiation could be fatal to any human passengers on-board. On the other hand, if a hydrogen-scoop spacecraft were designed to use conventional hydrogen-oxygen fueled engines, in contrast to advanced fusion engines, it would probably not be able to achieve such high speeds, but it could be built with existing technologies. It is worth noting that the primary reason for putting large numbers of people into space is to resolve the fundamental global problems of more and more people competing for fewer and fewer resources. Traveling at near the speed of light in order to reach distant galaxies is at best a secondary consideration. This is especially true when one considers that by the time such advanced technologies become available, biotechnology will probably have provided life spans that are virtually indefinite. This means the need to hurry along will no longer be such an important consideration. What follows is a general idea of how a state-of-the-art hydrogen-fueled spacecraft might be designed.

Hydrogen Starships

Assuming it is necessary to have a hydrogen scoop that is about 300 miles in diameter, the first consideration is that the scoop is such an enormous structure that it would probably be more practical to construct a fleet of four scoops that would be 75-miles in diameter, or eight scoops that would be about 35 or

40 miles in diameter. The hydrogen and other elements that would be collected could then be transported to the O'Neill-type cylindrical space colonies. Earth-like gravity (i.e., 1G) could be maintained by having the cylinders continually rotating. According to O'Neill, a typical space colony cylinder could be about four miles in diameter and about 16 miles long. Within such cylinders, large mountains and forests are possible, along with clouds, rain, lakes, rivers, grass, and animals. Agriculture would probably occur on other specialized cylinders that would be optimized for food-production purposes.

Space Colonies

The hydrogen and oxygen collected by fleets of such spacecraft would primarily be used for external propulsion and maneuvering. This is because all of the hydrogen-fueled vehicles and cities inside the biohabitat area of the space colony would be continually recycling the hydrogen used for energy purposes. Such spacecraft would not be dependent on either solar or nuclear energy, and as such, they could travel through space indefinitely, scooping up hydrogen, oxygen, and other elements like primitive microbes in the primordial sea.

Figure 5.47: Space Colony
The image was provided courtesy of the Space Studies Institute.

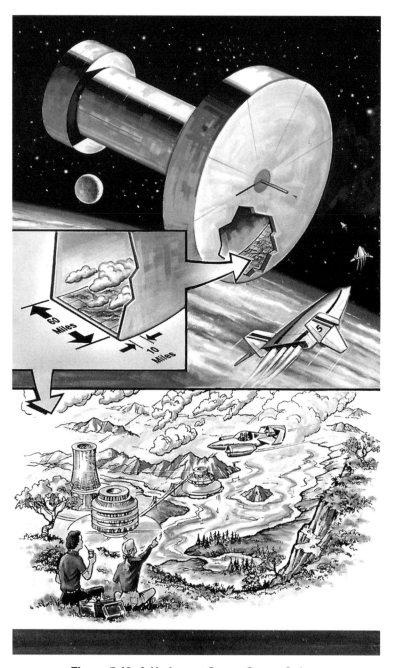

Figure 5.48: A Hydrogen-Scoop Space Colony

Figure 5.49: Arcology
A city without automobiles. The illustration is reprinted with permission from Massachusetts Institute of Technology at Cambridge.

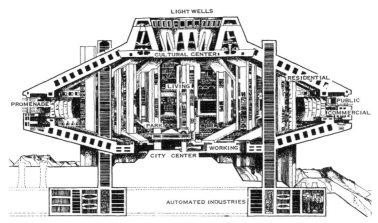

Population: 340,000 **Height:** 850 meters **Diameter:** 1,750meters

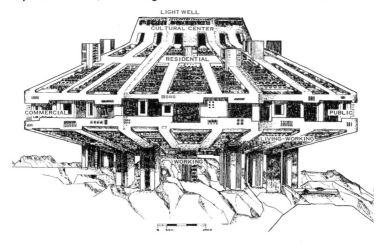

The Arcology illustration in Figure 5.49 is a megastructure designed by architect Paolo Soleri in Scottsdale, Arizona. Soleri refers to such structures as "arcologies" because they integrate architecture and ecology, by essentially eliminating the need for automobiles. With the lack of automobile-related support systems, such as roads and gas stations, the resulting urban area is reduced in size. Residential areas are located on the outer surfaces of the miles-high arcologies, while shopping, working and other indoor areas would be housed deep inside the structure.

Various prototypes and designs have been developed over the past thirty years by Paul Moller, who is a former professor of aeronautical engineering at the University of California at Davis. Moller now has his own firm, Moller International, which manufactures various automotive components, including the internal-combustion rotary engines that are at the heart of the Moller aircraft. A 2 or 4-seater Moller aircraft could be equipped with from three to six rotary engines that would be linked to an on-board computer to monitor engine functions. If all the engines should fail, the aircraft is to be equipped with a ballistically deployed parachute system, a shock-absorbing nose and fuselage, and airbags. At a cruising speed of about 300 mph, the Moller aircraft will be expected to average about 15 miles per gallon, which would provide it with a range of about 750 miles on its 50-gallon fuel tank, yet it is small enough to fit into a standard garage. The rotary engines are quieter than conventional aircraft engines. When hovering at 50 feet, a Moller aircraft will be expected to emit about 85 decibels of sound, which is less than one-third that of a Cessna 150 aircraft during takeoff. The rotary engines can burn virtually any fuel, including hydrogen, and when Moller-type aircraft are mass-produced, they will be expected to cost no more than existing luxury class automobiles.

Figure 5.50: The Moller Aerobot
The photograph was provided courtesy of Moller International.

Although having so many vehicles in the air could result in numerous collisions, it is anticipated that a highly reliable computer collision-avoidance system will eventually be a part of every aircraft. In addition, most urban traffic density is a result of millions of people driving to work on relatively narrow roadways that allow only a few lanes of traffic to pass at any given time. It is worth noting that if urban dwellers lived in Paolo Soleri-type arcologies instead of the unplanned chaos and urban sprawl that exists now in most urban areas, people would not need to use their vehicles in order to get to work, to school, or to go shopping. Although arcologies would be ideal urban environments in a futuristic hydrogen-fueled space colony, it is also possible to have urban and rural arcologies here on Earth. It is not really a question of technology, but political and social priorities in terms of determining what technologies should be developed. As the human and animal populations continue to increase, it will become increasingly necessary to have people living in space. As Buckminster Fuller noted, humans going into space is somewhat like a chicken coming out of its egg. This being the case, it is likely that the inhabitants of space colonies will want vast wilderness areas to simulate the Earth as much as possible.

Other Alternative Fuels

There are many other alternative transportation fuels that could be used other than hydrogen. These include reformulated gasoline, ethanol, methanol, natural gas (methane), ammonia or acetylene. When a switch to alternative or synthetic fuels is contemplated, it is especially important to examine all of the most viable options for three basic criteria:

1. Existing automotive and aerospace technology could be directed and optimized to utilize any of the alternative fuel options.

2. The cost of making an energy and industrial transition to any of the alternative fuels will be substantial.

3. Whatever substitute fuel is selected for gasoline, it ideally needs to be pollution-free and inexhaustible.

The primary reasons for selecting hydrogen instead of one of the many other alternative fuel options are that hydrogen is the most environmentally acceptable fuel; it is compatible with virtually all other energy sources; and it is completely renewable. The fact that hydrogen is inexhaustible is especially important given the projected exponential increases in global energy consumption that were discussed in Chapters 2 and 3. Equally important is the fact that once a transition to a solar-hydrogen energy system is made, there will never be a need to change to anything else. Moreover, hydrogen is one of the most efficient mediums for storing and utilizing the vast but intermittent solar energy resources. Howard Harrenstien, Dean of the School of Engineering & Environmental Design at the University of Miami, underscores the importance of hydrogen with the following observations:

"Many of the proposed new sources of primary energy, such as wind energy, ocean waves, ocean currents, ocean thermal differences, solar energy, and geothermal energy all suffer location disadvantages. They are generally not present in the right magnitude at the right place and at the right time for them to be directly converted to electricity and placed on line at that point. Even if they could be so converted, line losses imposed by transmission distances would place a heavy tax on efficiency. Enter hydrogen. Ocean thermal electric plants, which are located in the tropical seas, could generate it on location and store it at hydrostatic pressure on the sea bottom. Ocean tankers could collect the product at the proper time and deliver it to the world marketplace at great efficiencies in energy efficiency. Floating stable platforms, using photovoltaics and electrolysis to convert solar energy to hydrogen in mid-ocean, could also be placed 'on line.' Even bottom of the sea geothermal energy plants could be developed to produce hydrogen for these purposes. In sum, the hydrogen economy may hold the key to the integration of many new sources of energy into a common, environmentally acceptable synthetic fuel -- one which will allow us to conserve our precious fossil fuel reserves and, at the same time, develop a higher-level technology to advance the quality of life in this country and the world."

Hydrogen Production

Because hydrogen is the lightest of the 92 naturally occurring elements, if it is not chemically bonded to heavier atoms, such as carbon or oxygen, it will either drift up into the upper atmosphere where it will eventually react with oxygen to form water, or it will escape into space. It is for this reason that free hydrogen does not exist in significant quantities within the Earth's atmosphere. As a result, if hydrogen is to be used as a fuel or chemical feedstock, energy must be expended to separate the hydrogen atoms from the heavier atoms of carbon or oxygen.

The hydrogen that has been manufactured for industry is principally made by reacting natural gas with high temperature steam, thereby separating the hydrogen from the carbon. Gasoline is also a hydrocarbon molecule that is made up of eighteen hydrogen atoms that are attached to a chain of eight carbon atoms. In the case of natural gas, high temperature steam is used to separate the hydrogen from the carbon, and if the cost of the natural gas is $2 per million British thermal units (mmBtu), the cost of the gaseous hydrogen will be about $6.00 per mmBtu. If the hydrogen is liquefied, an additional $4.00 per mmBtu would need to be added, making the cost of liquid hydrogen produced by this method about $10.00/mmBtu. This corresponds to gasoline at the refinery costing about $1.22 per gallon.

Hydrogen can also be manufactured from coal-gasification facilities at a cost ranging from $8.00 to $12.00 per mmBtu, depending on the cost of coal and the method used to gasify it. But making hydrogen from nonrenewable fossil fuels does nothing to solve the basic problems of the diminishing resources, nor the serious environmental problems that result from using such resources for energy production. By contrast, if hydrogen is manufactured from water with wind-generated electricity, its cost will be roughly $25.00 per mmBtu, assuming the electricity costs are 4-cents/kWh. If the cost of electricity is reduced to 1 cent/kWh, the cost of hydrogen will be reduced to $15.00 per mmBtu, which is equivalent to gasoline costing about $1.85 at the refinery. Assuming the hydrogen is then liquefied, its cost would be increased to $19.00/mmBtu, which is equivalent to gasoline costing about $2.30 at the refinery. This relatively high cost has been, and continues to be, the major reason why hydrogen is not being produced on a large-scale by water electrolysis.

Manufacturing electrolytic hydrogen from water was initially anticipated by nuclear engineers who believed that nuclear-generated electricity would be inexpensive enough (too cheap to meter) to make hydrogen, but the high cost and unreliable nature of commercial reactors effectively eliminated the nuclear option. As such, a transition to a hydrogen energy system was put on hold. After the Arab oil embargo in 1973, there were long gas lines in the U.S. and the price of oil quadrupled. This prompted renewed research into alternative energy systems, including renewable energy technologies. As a result, significant progress has been made in developing the renewable energy technologies that could be mass-produced for large-scale hydrogen production. Some of the most cost-effective renewable energy options will be discussed at length in Chapter 6.

In addition to the more obvious solar and wind energy technologies, other renewable energy resources include the large quantities of human and agricultural wastes (including municipal sewage), paper, and other biomass materials that are accumulating in landfills. This is an important consideration because virtually all of the existing landfill sites in the U.S. will be filled to capacity before the year 2010. Moreover, generating hydrogen from such "waste" materials may turn out to be one of the least expensive methods of producing hydrogen -- and the resource is quite substantial. It has been estimated that in the U.S., roughly 14 quads of the annual 94 quad total energy requirement could be met from renewable biomass sources, which is roughly 14.8 percent of the total. With respect to sewage, it is both tragic and ironic that the vast quantities (literally billions of gallons per day) of human and animal waste that are being dumped into rivers and the oceans could -- and should -- be recycled to produce a renewable source of hydrogen. Researchers have shown that this can be accomplished either by utilizing the non-photosynthetic bacteria that live in the digestive tracts and wastes of humans and other animals, or by pyrolysis-gasification methods.

This is an important consideration because marine biologists are already warning that the microbial food-chains in the oceans are in serious trouble and may be about ready to collapse because of the enormous levels of sewage and other toxic wastes that are being dumped daily. Since roughly 60 percent of the oxygen in the Earth's atmosphere comes from the microorganisms that live in the oceans, having them suddenly disappear will be catastrophic for most oxygen-breathing organisms. This problem is compounded by the fact that most of the oxygen-producing microbes live primarily on the continental shelves where there is sufficient light for the microbes to operate their photosynthetic processes. Unfortunately, these continental shelves are the very areas where most of the raw sewage and other toxic chemicals are being dumped.

As such, the national security interests of the nation are now dependent on solving the human, animal and toxic waste problems. It should logically follow that such environmental concerns should have similar budgets as do the weapons programs. If billions of dollars were spent to develop and deploy advanced sewage treatment systems then the billions of gallons of raw sewage that is being dumped into the oceans every year could instead be transformed into relatively low-cost hydrogen. Although high-temperature nuclear-fusion reactors may some day be practical as renewable sources of energy for hydrogen production, they are at present only a theoretical option. Typically, over 100 million degree Fahrenheit temperatures are required for nuclear fusion to occur, and as a result, such highly sophisticated technologies are not expected to be commercially viable for many decades, if not centuries. Before energy technologies can be considered economically viable they must first be technically viable; and high-temperature fusion reactor systems are a long way from being either. Such energy options cannot be realistically expected to produce any significant quantities of hydrogen or electricity before the existing 10 to 20-year supply of known U.S. oil reserves are substantially exhausted. There is always the possibility that new and innovative nuclear energy systems could be developed for safe and economical energy production purposes, but long-term energy decisions should not be predicated on nonexistent or unproven technologies.

Conclusions

An industrial transition from a petroleum economy to a hydrogen economy would resolve many of the most serious global energy and environmental problems. The problems associated with fossil fuel and nuclear fission energy systems underscore the importance of developing renewable energy options that do not pose such long-term and unknown environmental and economic risks. When one undertakes a careful review of the most viable renewable energy options that could generate enough hydrogen to displace the use of fossil and nuclear fuels, one is inevitably reduced to the relatively simple technology options that are in one way or another associated with solar or biological energy processes and resources.

It is worth noting that biological organisms have been successfully utilizing a solar-hydrogen energy system on a global scale for over 3.5 billion years. Because solar-hydrogen technologies can be mass-produced in automotive-type industries, they will be able to produce energy that is economically competitive with that generated from either fossil or nuclear energy systems, while providing significant private-sector employment and economic stability for many decades.

Most importantly, such an energy transition will allow civilization's millions of industrial machines to function in relative harmony with the Earth's natural biological life support systems. Fortunately, there are many viable solar options (which will be discussed in Chapter 6) that can be mass-produced for large-scale hydrogen production. Because such systems are no more complex to manufacture than automobiles, airplanes or ships, they could be rapidly implemented on a global-scale with existing technology. Willis M. Hawkins, who is a senior technical advisor and former Board Member of Lockheed Aircraft Corporation, provides a note of encouragement when he writes:

"I certainly hope that those who recognize the advantages of hydrogen for industrial and home consumption will be pressing just as hard on the infrastructure serving those customers as we will be for aircraft. We hope that the pioneers and developers in other fields will challenge us or even help us to lead in converting the world to hydrogen."

For a more detailed historical discussion of the hydrogen energy developments, read *The Forever Fuel: The Story of Hydrogen* (Westview Press, Boulder, Colorado), written by Peter Hoffmann, a former deputy bureau chief of McGraw-Hill World News in Bonn, Germany, and editor of a hydrogen newsletter. Hoffmann's book provides an excellent overview of the discovery, development and end-use of hydrogen as an energy medium.

The most extensive single source of technical information on hydrogen energy is provided in the *International Journal of Hydrogen Energy* (published by Pergamon Press), which is the official publication of the International Association for Hydrogen Energy (IAHE). The IAHE is a peer-review society involving some of the world's most distinguished scientists and engineers from within industry, education and government. It is a technical brain trust that has representatives from over 80 countries who have been carefully assembling the necessary engineering and chemical information that will be required when the world decides to make an industrial transition to renewable resources a reality. T. Nejat Veziroglu, Ph.D., who is an engineering professor at the University of Miami, Coral Gables, Florida, is president of the IAHE. He is also Editor-in-Chief of the IAHE Journal, and is one of its original founding fathers.

As Veziroglu has written:

"We do not have to subject the biosphere and life on this planet to the deadly effects of fossil fuels. The answer is to establish the environmentally most compatible, clean and renewable energy system, the Hydrogen Energy System. What we need now is the governmental decisions to convert to the new energy system in an expeditious and prudent manner. When the subject is brought up with government officials, their answer is 'Hydrogen sounds good, but just now it is more expensive than petroleum. Let the free market forces decide what the energy system will be.' The answer is fine, if the rules of competition are fair. At the present time, they are not. The rules favor the lower production cost products, irrespective of their environmental effects. We need new and fair laws which take into account environmental damage, as well as production costs...Once such a principle is legislated, then no energy company will produce and sell petroleum for fuel, but will begin to manufacture hydrogen, since it will be by far the cheapest fuel."

The fact that the existing free market pricing system does not take into account the many environmental costs that result from using fossil fuels is only part of the problem. Consider the hundreds of billions taxpayer's dollars that have been expended on building and maintaining the military forces in the Middle East. Clearly, the U.S. forces in the Middle East are deployed to protect the oil supply lines, and if the oil companies were required to pass these costs along to the consumer at the gas pump, gasoline would have never been able to economically compete with hydrogen. Thus, if a "fair accounting" system were implemented, hydrogen would be the least expensive fuel in a free market economy. Carl Sagan is quoted in the beginning of this chapter documenting the fact that we as human beings are essentially made up of "starstuff." It would, therefore, seem appropriate to conclude this chapter on hydrogen by expanding on Sagan's explanation of the origin and evolution of hydrogen and the other elements in the known universe. The following passage is quoted from Sagan's remarkable *Cosmos* television series that was originally aired on the Public Broadcasting System (PBS):

"Some 15 billion years ago, our universe began with the mightiest explosion of all time. The universe expanded, cooled and darkened. Energy condensed into matter, mostly hydrogen atoms, and these atoms accumulated into vast clouds rushing away from each other, that would one day become the galaxies. Within these galaxies, the first generation of stars was born, kindling the energy hidden in matter, flooding the Cosmos with light. Hydrogen atoms had made suns and starlight. There were in those times no planets to receive the light and no living creatures to admire the radiance of the heavens. But deep in the stellar furnaces, nuclear fusion was creating the heavier atoms; carbon and oxygen, silicon and iron. These elements, the ash left by hydrogen, were the raw materials from which planets and life would later arise. At first, the heavier elements were trapped in the hearts of the stars, but massive stars soon exhausted their fuel, and in their death-throes returned most of their substance back into space. Interstellar gas became enriched with heavy elements. In the Milky Way galaxy, the matter of the Cosmos was recycled into new generations of stars, now rich in heavy atoms, a legacy from their stellar ancestors. And in the cold of interstellar space, great turbulent clouds were gathered by gravity and stirred by starlight.

In their depths, the heavy atoms condensed into grains of rocky dust and ice, complex carbon-based molecules. In accordance with the laws of physics and chemistry, hydrogen atoms had brought forth the stuff of life. Collectives of organic molecules evolved into one-celled organisms. These produced multi-celled colonies, when various parts became specialized organs. Some colonies attached themselves to the sea floor. Others swam freely. Eyes evolved, and now the Cosmos could see. Living things moved on to colonize the land. Reptiles held sway for a time, but they gave way to small warm-blooded creatures with bigger brains who developed a dexterity and curiosity about their environment. They learned to use tools and fire and language.

Starstuff, the ash of stellar alchemy, had emerged into consciousness. These are some of the things that hydrogen atoms do, given 15 billion years of cosmic evolution. It has the sound of epic myth, but it's simply a description of the evolution of the Cosmos, as revealed by science in our time."

Figure 6.1: Wind Energy Systems
To date, wind energy conversion systems have been the most cost-effective of any of the renewable energy technologies.

CHAPTER 6

RENEWABLE ENERGY
TECHNOLOGIES

In order to displace the use of fossil and nuclear fuels in the U.S. with renewable hydrogen fuel, the size of the electricity production system in the U.S. (and most other industrialized countries) will have to be increased by a factor of three. As a result, the major infrastructure issue that needs to be resolved has to do with the need for a greatly expanded electricity production capability. In 1998, the U.S. generated about 73 quadrillion (quads) Btu, but consumed about 94 quadrillion (quads) Btu of energy, of which roughly 33 quads was used to generate electricity. In order to provide the total U.S. energy requirements with electrolytic hydrogen production systems, roughly 10 million megawatts of electrical generation would need to manufactured and installed. Given the scale and scope of such an effort, only the renewable energy technologies have the potential to be mass-produced for such a task. As a result, some of the most important questions are: *What technologies will be the most cost-effective for large-scale hydrogen production? How much will this project cost and how long will it take?* There is a wide range of renewable energy technologies that have been under investigation for the past several decades. Some of the more promising renewable energy technology options that will be discussed in this chapter that could be mass-produced for large-scale electricity and/or hydrogen production include photovoltaic cells, solar engine systems, wind energy and ocean thermal energy conversion (OTEC) systems that are able to utilize the vast amounts of solar heat that is stored near the surface of the oceans.

Energy Economics

In evaluating the various renewable energy options, or any energy technology for that matter, the primary concern is economics. Many people support the use of renewable energy technologies, but if the energy produced from these systems is three or four times more expensive than energy from fossil or nuclear fuels, most of those same people will use the latter and not the former. When calculating energy costs, however, one can come up with very different numbers depending on what variables are used in the equation. In the case of electricity, for example, the costs are generally referred to as "cents per kilowatt hour" (cents/kWh). This number, if calculated accurately, must include a system's capital and construction costs, and all operation and maintenance costs, including fuel costs, if any, as well as financial costs, including costs of money, insurance and taxes. Wholesale electricity costs are typically in the range of 3 cents/kWh during off-peak hours of the day or night, whereas wholesale electricity costs during peaking hours can often be over 15 cents per kWh. Electricity generated from hydropower projects is generally estimated to cost 1 to 2 cents/kWh, and is often referred to as some of the least expensive electricity available. Yet, in virtually all of the hydropower projects in the U.S., the federal government paid for the capital costs of building the dams, with low-interest long-term financing. If the costs of hydropower projects include the capital costs and conventional financing terms, the true cost of electricity is in the range of 8 to 10 cents/kWh.

Utility executives have testified at rate hearings that electricity generated by nuclear fission plants costs about one cent/kWh -- without explaining that the one-cent number only refers to the fuel costs, and even these nuclear fuel costs are subsidized by the federal government. The true costs of nuclear-generated electricity are, in fact, unknown, because no one has been able to accurately calculate how much it will cost to store the radioactive wastes over a period of several hundred thousand years. As a result, these huge numbers are simply left out of the equation.

Renewable energy technologies are generally viewed by most energy and utility analysts as not being economically competitive with conventional fossil fuel or nuclear facilities. However, this assumption is both inappropriate and to a certain extent, inaccurate. The assumption is inappropriate because fossil fuels are nonrenewable, which means their prices are inherently unstable. While the surplus of oil reduced its cost for the decade of the 1980s and the 1990s, it is only a question of time before the gas lines and high prices return. It is well to remember that the world currently produces only one barrel of oil for every four it consumes. The time to fix the roof is before it begins to rain. To use the current price of fossil fuels in terms of long-range energy policy formation is like trying to drive a car while continuing to look in the rear view mirror. One does not realize one is in trouble until after an accident occurs.

External Costs

In addition to the direct costs associated with fossil fuel and nuclear energy systems, there are many "external costs" that are not factored into the energy cost equation. Such external costs include environmental damage, medical and military costs. If these costs were factored into energy cost equations, the renewable energy technologies would be the least expensive to use. In addition, if the renewable energy technologies are mass-produced on a scale to displace fossil fuel and nuclear energy systems, their capital and overall operating costs will continue to be reduced in the future. External costs are not currently factored into the energy cost equations because such costs are staggering and in many cases, difficult, if not impossible, to accurately calculate. While calculating U.S. military costs in the Middle East is relatively easy, it is much more difficult to estimate the billions of dollars spent each year on medical costs that are related to having millions of people live in highly polluted cities. Even without considering the environmental costs, there are many renewable energy technologies that could produce electricity and/or hydrogen at prices that are competitive with fossil-fueled or nuclear facilities. Unfortunately, few people are aware of the most cost-effective renewable energy technologies that could resolve many of the most serious environmental problems that threaten to make the Earth uninhabitable. It is interesting to note that the renewable energy technology option that has received the vast bulk of media and press exposure, as well as most of the U.S. government research dollars, happens to be one of the most expensive: photovoltaic cells.

Photovoltaic Cells

Photovoltaic (PV) solar cells are semiconductor devices that are able to convert sunlight directly into electricity with no moving parts. Because PV cells have successfully provided electricity for space vehicles for many decades, as well as a wide range of consumer electronic devices such as calculators and watches, they have received most of the attention in the media. Indeed, for most people, when solar-electric energy systems are mentioned, they assume one is referring to PV systems.

The "photoelectric effect" was first observed in 1881 by the German physicist, Heinrich Hertz. He discovered that light (which consists of the visible wavelengths of electromagnetic radiation) could displace electrons from certain metals. A PV cell is typically made of a material such as silicon, which acts as both an electrical insulator and a conductor. The silicon strips are coated with other materials, such as boron, in order to produce a positive electrical layer, which then interacts with the underlying negatively charged silicon layer. This positive-negative area is referred to as the "p-n junction," and when a photon of sufficient energy impacts an electron in the silicon, the electron moves across the p-n junction, and in doing so, produces a flow of direct-current (DC) electricity. Silicon-based PV system efficiencies have typically been in the range of 10 to 12 percent.

While PV costs have come down dramatically in recent years, installed PV systems still cost roughly $10,000 per kW, compared to fossil fueled combustion turbines that can be installed for about $500 per kW. Not surprisingly, this high cost has prohibited the use of PV systems for base load energy production purposes. Although the cost of PV systems has been reduced from roughly $50,000 per kW in 1974, their costs will need to be reduced to $1,000 per kW if they are to be competitive. Electricity generated from natural gas-fueled plants costs on average from 2 to 6 cents per kWh, whereas electricity costs from PV systems are roughly 10 times higher. Also, the efficiency of PV cells is reduced in the summer when the utilities need the power the most, and PV systems also lose between one and two percent of their electrical output annually. While it is reasonable to assume that the cost of PV systems will continue to be reduced in the future, their electricity costs are presently far too expensive for the generation of either electricity or hydrogen.

Figure 6.2: A Photovoltaic Array
Photovoltaic solar cells provide electricity with no moving parts for a
residence in the Arizona desert.

Figure 6.3: A Centralized PV Array
Megawatt-scale photovoltaic systems can be created by connecting the
modularized panels to each other.

Although a PV cell may only be 10 percent efficient in converting sunlight into electricity, it is well to remember that photosynthetic green plants are only about one percent efficient and they have been successfully operating on a global scale for over three billion years.   Moreover, comparing solar efficiencies with the efficiencies of conventional power plants is like comparing apples with oranges. What is appropriate is to compare the efficiencies of other solar technologies that have been developed in recent years, because while PV systems have received the vast majority of solar research funds, they are still one of the most expensive solar technology options available. It is only a question of time before there will be an efficient and cost-effective PV system developed.   However, if existing PV systems were the only solar option available, the industrial transition to renewable resources would not be economically viable. Fortunately, there are other economical solar technology options that have been developed in recent years. Three of the most promising solar technologies are solar engine systems, wind energy conversion systems, and ocean thermal energy conversion (OTEC) systems.

Solar Engine Systems

Solar engine systems, which look similar to a satellite or radar dish, have held the world's efficiency record for converting solar energy into grid-quality electricity since 1984. Solar engine systems are roughly three-times as efficient as PV systems, but their overall operation and maintenance requirements have thus far been significantly higher than PV systems that have essentially no moving parts. As a result, in remote off-grid applications where a relatively small amount of power (i.e., one to five kW) is needed for a home or facility, PV systems are the most practical option. However, if larger amounts of power are needed for utility-scale grid-connected systems (i.e., from 1 to 100 MW), solar engine systems have the clear advantage. Solar engine systems typically integrate a heat engine, such as a Stirling or Brayton-cycle electrical power conversion unit (PCU), with a point-focus solar concentrator system. The solar concentrator component of a solar engine system acts like a large magnifying glass, which concentrates the incoming solar energy onto a target area where the PCU is located.

Figure 6.4: Solar Engine Systems
Two prototype solar engine systems are operating "on Sun" at Edwards Air Force Base in Southern California. These two systems were initially developed and field tested in the 1980s by the U.S. Department of Energy in cooperation with the Jet Propulsion Laboratory.

It is worth noting that two of the engines used in the JPL tests pictured earlier in this Chapter in Figure 6.4 were not initially developed for the prototype solar engine gensets, but for automobiles. Solar engine systems are one of the most potentially cost-effective solar options because they can utilize highly efficient Stirling-cycle engines, which can be mass-produced with existing automotive technology. Although solar engine systems are in some respects like technological trees (in that they will use sunlight and water to make hydrogen); from a manufacturing perspective, they are more like automobiles. The individual sections of solar concentrator systems are made of non-strategic materials, such as glass and steel, and they would be about the same size as the hoods for automobiles. Individual solar gensets can be connected together for large-scale electricity or hydrogen production; and unlike nuclear or fossil fuel systems, they would produce electricity with only minimal water requirements and without the production of any toxic wastes.

Solar engine systems are expected to be less expensive by weight than typical automobiles because they have a simpler design and they will not need to be changed annually for cosmetic reasons to enhance sales volumes. The fact that solar gensets are similar to automobiles from a manufacturing perspective means economic and engineering assumptions can be directly extrapolated from the experience of the automotive industries. Moreover, because solar gensets can be mass-produced, the electricity they generate has the potential to be less expensive than the electricity generated by fossil fuel or nuclear facilities. Calculations indicate that if 12 twenty-five kW solar gensets are installed per acre, roughly 35 million acres (i.e., less than ten percent of the desert areas in the American Southwest) would be enough land to make the U.S. energy independent of the Earth's remaining fossil fuel and uranium reserves.

Early Solar Stirling Systems

The first known solar engine system was designed and built right after the Civil War in the U.S. by John Ericsson, a Swedish-American engineer and inventor. Ericsson is perhaps best known for his design of the ironclad Monitor warship that was built for the U.S. Navy during the Civil War.

Ericsson also designed and developed a number of other noteworthy technologies, including the first screw-propeller for steamships, the first revolving gun turret, and the first ship where the machinery in the ship was safely moved below the water line. Ericsson was fascinated by the potential of solar-powered engines, and by 1875, he had developed seven different types of what he referred to as "Sun motors." Ericsson's first Sun motors had incorporated steam engines for power production, but because of the continual valve failures he experienced, his later designs utilized closed-cycle Stirling engines. In a letter sent to a friend and business associate in 1873, Ericsson explained that in contrast to his earlier models, his new Stirling engine design was absolutely reliable. By 1880, he had developed his Stirling Sun motor system into a conventionally fueled water pump, and thousands of the machines were used worldwide.

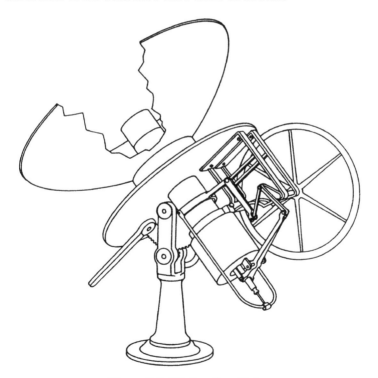

Figure 6.5: Stirling Sun Motor
A drawing of a Stirling "Sun Motor" designed and built by John Ericsson in 1872.

The solar concentrator system pictured in Figures 6.6 and 6.7 is a state-of-the-art design that was developed by McDonnell Douglas in the early 1980s. The Stirling engine power conversion unit (PCU) was developed and manufactured by Kockums, one of the largest defense contractors in Sweden that spent over $250 million in the development of Stirling-cycle PCUs for use in powering non-nuclear attack submarines. Although the McDonnell Douglas/Kockums system was well engineered, both companies elected to discontinue their involvement in the development of solar engine systems in 1986. McDonnell Douglas sold their solar Stirling technology to Southern California Edison (SCE), which continued to test the equipment until 1988 when it also discontinued its solar thermal research program.

SCE did not terminate the solar engine program because the technology failed to operate successfully. Rather, research engineers from SCE concluded that both from a technical and economic perspective, the solar Stirling systems were one of the most attractive solar energy systems currently available. Rather, SCE had received new guidelines from the California Public Utilities Commission that discouraged utilities from being involved in solar research and development. In addition, SCE was not planning to build any new electrical generating facilities for the foreseeable future. This was, in part, because utility deregulation was being anticipated at a time when growing numbers of large industrial users were increasingly generating their own power on-site with natural gas-fueled cogeneration systems.

The McDonnell Douglas solar concentrator technology was acquired from SCE in 1996 by Stirling Energy Systems, Inc. (SES), a Phoenix, Arizona-based systems integration firm that also acquired the exclusive U.S. manufacturing rights to the Kockums solar Stirling engine that was integrated with the system. SES was able to get McDonnell Douglas and Kockums re-engaged in the commercialization of the solar engine system in 1997, and this SES team, led by McDonnell Douglas, secured a "Dish Engine Critical Component" (DECC) contract with the U.S. Department of Energy (DOE) in 1998 to do endurance and reliability testing for the system. At the end of Phase One of the DECC contract, the technical program manager from Sandia National Laboratories reported the SES/Boeing solar Stirling program was one of the most successful solar demonstration programs in DOE's history, in terms of reliability and performance.

Figure 6.6: 7-Generations of Concentrator Development

McDonnell Douglas secured over $100 million in contracts to develop solar heliostats for central receiver systems before spending over $10 million of its own company funds to develop the solar concentrator system that would be used to operate with the Kockums Stirling PCU.

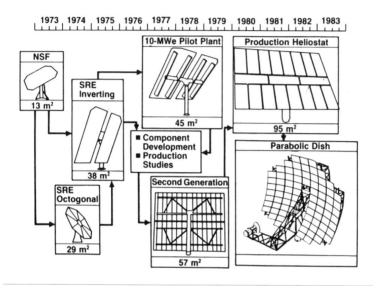

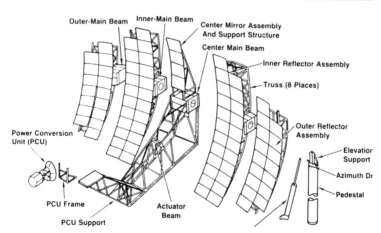

Figure 6.7: Major Subassemblies

The McDonnell Douglas "balanced" solar concentrator system offsets the weight of the solar concentrator with the weight of the Stirling PCU. The concentrator was developed in five major subassemblies that could be transported by truck and assembled in the field in about four hours.

Figure 6.8: PCU Maintenance
The McDonnell Douglas solar concentrator was engineered with a slot down the middle of the system, which allows the system to "kneel" so that the solar technicians can have easy access to maintain the PCU.

Figure 6.9: Boeing/SES Solar Test Site
Two of the SES solar concentrator systems have been operating at the Boeing facility in Huntington Beach, California since 1984. The optical performance of the systems has been unchanged during this period.

Figure 6.10: SES/Boeing Solar Stirling System
This 25 kW solar Stirling system initially developed by McDonnell Douglas is now being commercialized by Boeing and Stirling Energy Systems. A 10 MW central receiver tower is glowing in the background.

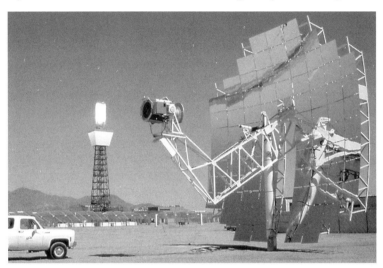

Figure 6.11: Solar Fireball
A reflection of the target area shows where the solar concentrator system focuses the Sun's heat to power the PCU. Each one of the 82 mirror "facets" is individually aligned for optimum system performance.

DOE has also been funding a solar Stirling system that has been developed and field tested by Science Applications International Corporation (SAIC) and Stirling Thermal Motors (STM), located in Ann Arbor, Michigan. The 25 kW "Sundish" system, pictured in Figure 6.12, is now being field tested at a number of utility test sites in the American Southwest, including Arizona Public Service Company and Salt River Project in Phoenix, Arizona. The STM Stirling engine is similar to the Kockums Stirling engine in its size and output, and because of their "external combustion" characteristics, both Stirling engines have the potential to be used in non-solar applications with a wide-range of chemical fuels, such as natural gas or gasoline. Stirling PCUs are also able to use renewable biomass-sourced fuels, such as woodchips or corncobs, or the methane that is generated from landfills and wastewater treatment plants. This ability to use chemical fuels means the Stirling PCU can provide "dispatchable" power 24 hours per day.

Figure 6.12: STM SunDish System
The photograph was provided courtesy of Stirling Thermal Motors.

Figure 6.13: Solar Stirling Power Plant
A state-of-the-art solar Stirling system initially developed by McDonnell Douglas that is now being commercialized by SES and Boeing. The photograph was provided courtesy of Stirling Energy Systems.

Figure 6.14: Solar Dish Forest
The image was provided courtesy of U.S. Naval Research Laboratory.

Figure 6.14 is a painting by Pierre Mion that provides a prospective view of what a large "forest" of solar engine systems might look like in a typical desert region. The solar gensets are like "technological trees" because they will be able to use sunlight to break water down into hydrogen and oxygen.

Stirling-Cycle Engines

The Stirling-cycle engine was invented in 1816 by Robert Stirling, a Scottish minister, as well as an esteemed scientist and engineer. Stirling's interest in developing a more efficient engine was stimulated by the fact that steam engines at the time were made with relatively poor alloys, and as a result, they had a habit of blowing up and killing and/or injuring anyone who might be nearby. Stirling's patent was titled "Improvements for diminishing the consumption of fuel, and in particular an engine capable of being applied to the moving machinery on a principle entirely new." Stirling's technical innovation was to reuse heat that would otherwise be wasted in conventional engines. What he created in the process was one of the most efficient thermodynamic mechanical cycles ever developed: the Stirling-cycle.

Conventional gasoline-fueled engines use an "Otto cycle" homogeneous-charge combustion process that was developed in 1870 by Nikolaus Otto. Diesel-fueled engines, on the other hand, utilize a more efficient stratified-charge combustion process that was developed in 1893 by the German engineer, Rudolf Diesel. In the Otto-cycle engine, heat input, a result of spark ignition of the fuel is provided at a constant volume, while in a Diesel engine, compression ignition occurs under constant pressure. Both the Otto and Diesel-cycle engines are internal combustion engines that ignite the fuel inside the cylinder.

By contrast, Stirling-cycle engines utilize an "external" combustion process, whereby the source of heat occurs outside the cylinder of the engine. The external heat is transferred through a heat exchange material to a working gas (typically hydrogen or helium) that is sealed inside the Stirling engine. As the working gas is heated, it expands, and in so doing, the pressure wave causes the piston to move. This is why a Stirling-cycle engine is omnivorous (i.e., it can use virtually any form of energy or combustion fuel that creates heat, including solid biomass fuels as well as concentrated solar energy). A cutaway of a chemically fueled Stirling engine developed in Sweden by Kockums is shown in Figure 6.17. Note that the "external combustion" fuel igniter near the top right hand corner of the engine is external to the interior piston assembly, thus none of the corrosive by-products of combustion, such as organic acids or carbon deposits, are introduced into the interior components of the engine.

Figure 6.15: Robert Stirling
Robert Stirling, "father" of the "external combustion" engine, which is one of the most efficient thermodynamic cycles: *The Stirling cycle.*

Figure 6.16 Automotive Stirling Engine
This four cylinder 4-95 Stirling engine, which was optimized for an automobile, was developed by Kockums, in cooperation with Volvo, Ricardo of England, and Ford. NASA has also worked extensively with Kockums to improve the performance and cost of the automotive Stirling engine.

This lack of corrosion is significant because it allows a Stirling engine to operate by a factor of 10 longer than conventional internal combustion engines. In conventional internal combustion engines, the fuel is injected into the interior of the cylinder and ignited with a spark plug, or in the case of a Diesel engine, compression ignition occurs. In either case, the corrosive by-products of combustion, such as organic acids and carbon deposits, contaminate the engine oil and cause corrosion and wear to bearing surfaces. An external combustion engine also has greatly reduced noise levels and environmental emissions. These considerations make Stirling engines ideally suited for power generation applications.

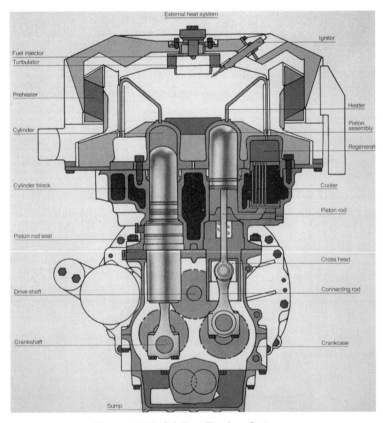

Figure 6.17: Stirling Engine Cutaway
The image was provided courtesy of Kockums
& Stirling Energy Systems.

Figure 6.18: Stirling-Powered Attack Submarine
NASA has worked extensively with Kockums, one of the largest defense contractors in Sweden, to refine their Stirling engines for automotive and space applications. Kockums has spent over $250 million over the past 20 years to develop Stirling PCUs, which are used by the navies of Sweden and Japan for non-nuclear attack submarines like the one pictured below.

Figure 6.19: Stirling PCU
The Stirling PCUs allow the submarine to increase its underwater operation time by a factor of seven compared to Diesel electric propulsion systems. The photograph was provided courtesy of Kockums AB.

Figure 6.20: Solar Stirling Engine Cutaway

Note that the solarized Stirling engine is integrated with a solar receiver in place of a conventional burner system. The image was provided courtesy of Kockums and Stirling Energy Systems.

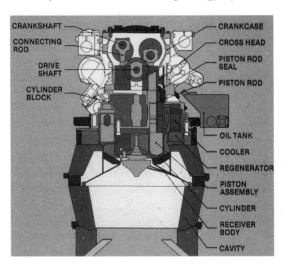

Figure 6.21: Solar Stirling PCU Without Receiver Enclosure

With the solar receiver assembly removed, it is possible to see the spherical heater head tubes that are shown at the front of the engine where the hydrogen gas is circulated. The image was provided courtesy of Kockums and Stirling Energy Systems.

Figure 6.22: Solar Stirling PCU
The solar Stirling power conversion unit (PCU) is shown with the solar receiver assembly attached. The hydrogen gas bottle is located on top of the 25 kW electrical generator that is directly coupled to the engine drive shaft. The image was provided courtesy of Kockums.

Figure 6.23: The Stirling PCU "On Sun"
The octagonal structure that surrounds the Stirling engine is an automotive-type radiator filled with water and antifreeze. The image was provided courtesy of Kockums.

Figure 6.24: Stirling Heater Head

The heater head of the Stirling engine is made up of a group of inconel tubes that carry the hydrogen working gas. As the hydrogen gas is heated, it expands against a piston assembly that is connected to a crankshaft, which can, in turn, be connected to an electrical generator. Heater head operating temperatures are about 700 to 800 degrees C.

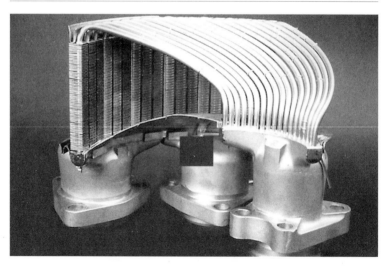

Figure 6.25: Heater Head Quadrant

One of the four quadrants of a typical Stirling heater head is shown. When all four quadrants of the heater head are braised in a furnace, they are forged into a single circular assembly that is pictured in Figure 6.24. Both of the photographs were provided courtesy of Kockums.

Figure 6.26: Mobile Stirling PCU

Stirling electrical generator-set systems are ideally suited for mobile applications.because they can operate with virtually any fuel, their "external combustion" process makes them exceptionally quiet while they are operating, and it also eliminates carbon deposits and organic acids from being introduced into the interior components of the engine.

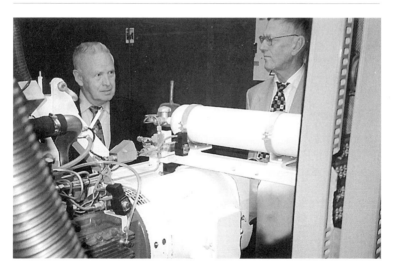

Figure 6.27: Stirling PCU on Bench

Ken Stone (left), the solar Stirling program manager for Boeing reviews a solar Stirling engine on the bench with Christer Bratt, one of the principal engineers at Kockums who has been developing Stirling engines for both non-nuclear attack submarines and solar concentrator systems.

Early Stirling engines, built in the late 1800s and early 1900s, were considered to be reliable, but also relatively heavy, giving them a low power-to-weight ratio. Internal combustion engines have a power density of about 4 to 5 pounds per horse-power in contrast to about 13 pounds for the early Stirling engines. Early Stirling engines were also relatively expensive be-cause the external combustion required the heat to be transferred through heat exchangers and ducts that required more expensive nickel-based alloys. As a result, while major automobile manu-facturers like General Motors and Ford continued to research Stirling engine designs, they elected to utilize Otto and Diesel en-gines for their automotive vehicles. Although Otto and Diesel cycle engines are well suited for transportation vehicles, they are unusable with solar engine systems because of their internal combustion characteristics. A solar engine system requires an engine that operates on external heat, such as a Stirling, Rankine or a Brayton (gas turbine) cycle design.

After the Arab oil embargo of 1973, there was a serious attempt in the U.S. to investigate the most cost-effective solar technologies. As a result, several dish test-bed systems were constructed for NASA and the Department of Energy with all three of the engine configurations. It became clear from the field tests that the Stirling PCUs were the most promising in terms of efficiency, long life and cost.

There are two basic types of Stirling engines: kinematic and free-piston. Kinematic configurations are similar to conventional internal combustion engines in that they have a crankshaft or swashplate driven by pistons. Free-piston Stirling engines, on the other hand, do not have a crankshaft, valves or any physically connected moving parts. They have only two moving parts whose motion is determined by their respective masses, as well as gas bearings (i.e., there is no physical connection between moving members). The primary components of a free-piston Stirling engine include a displacer and power piston that are sup-ported by hydrostatic gas bearings with noncontacting clearance seals. The integration of the free-piston engine with a linear alter-nator (for electricity production) is schematically shown in Figure 6.28. When heat is applied, the working gas expands, causing the piston to move back and forth. Because of the free-piston engine's simplicity, they have the potential for significantly reduced maintenance requirements.

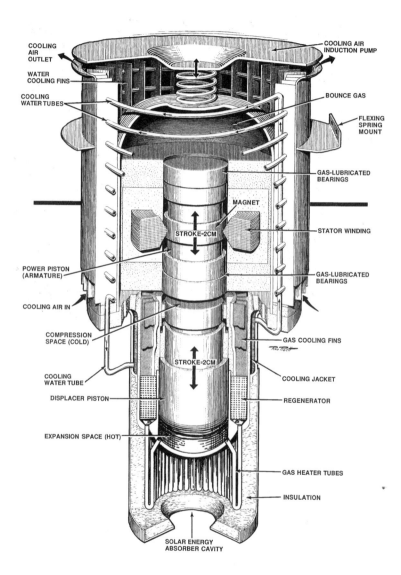

Figure 6.28: A Free Piston Stirling Engine
The above cutaway is of a free-piston Stirling engine integrated with a linear electrical alternator. The system was developed by professor William T. Beale, College of Engineering, Ohio University. Reprinted from Popular Science with permission © 1985 Times Mirror Magazine.

Figure 6.29: Free Piston Design Team

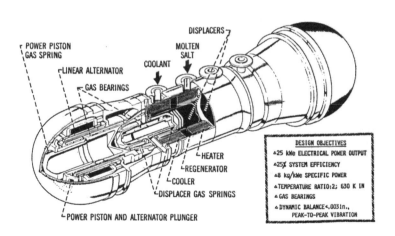

Figure 6.30: NASA Stirling PCU
The dual cylinder free-piston Stirling engine pictured above was developed for NASA by Mechanical Technologies Inc. in Latham New York. The engine has a 25 kW power output. Image courtesy of MTI.

Free-piston Stirling engines are especially quiet, hermetically sealed and literally welded shut, making them airtight. This means it would be impossible for dust or other particles to enter the interior of the free-piston Stirling engine. Most importantly, because of their simplicity, the free-piston Stirling engines may be the least expensive to manufacture. One of the most advanced concepts for a free-piston solar Stirling system has been proposed by Glendon Benson, Ph.D. and his engineering research and development colleagues who prepared a paper, "An Advanced 15 kW Solar Powered Free-Piston Stirling Engine." Their paper was delivered in 1980 at the 15th Intersociety Energy Conversion Engineering Conference held in Seattle, Washington, and later published by the American Institute of Aeronautics and Astronautics.

It was Benson and his colleagues who first proposed using solar Stirling engine systems for large-scale hydrogen production. Hydrogen would ideally be generated with a high-temperature system that has been under development by investigators at Brookhaven National Laboratory and Westinghouse Corporation (now Siemens). Such a system could be integrated into the solar receiver that contains the fireball of hot air, and when optimized, overall system efficiencies in excess of 60 percent were predicted. This was in contrast to the maximum efficiency of 28 percent for silicon-based PV cells; but despite this significant announcement, the paper by Benson and his colleagues went unnoticed by the media and international energy engineering community.

Although one might be tempted to conclude that the optimum solar technology would integrate a solar concentrator with a free-piston Stirling engine, free-piston Stirling engines have thus far been shown to only be about half as efficient as their kinematic cousins. As a result, the Stirling engines that are being used in the solar engine systems being developed by the SES/Boeing team and the STM/SAIC team both utilize kinematic Stirling engine configurations. In the case of the STM Stirling engine, the conventional crankcase has been replaced by a "swashplate" subsystem for converting the power of the connecting rods into a rotating electrical generator, but the STM engine is still classified as a kinematic engine. While free-piston engine configurations are promising, it is well to remember that there are millions of engines manufactured every year, and they are all kinematic machines.

Manufacturing Costs

Stirling engine manufacturing cost studies have been completed by a number of automotive manufacturers, including Volvo, Hercules Engine Company, Cummins Diesel Engine Company and John Deere, in cooperation with NASA and MTI in Latham, New York to analyze Stirling engine costs. Recent studies have also been undertaken by SES/Boeing and the STM/SAIC solar Stirling teams. Based on these studies, it is reasonable to expect the cost of a Stirling engine to be comparable to a Diesel engine, assuming the same number of units were produced annually. A kinematic Stirling engine with a 40,000-hour life would be expected to cost about $100 per installed kW, assuming 25,000 units were produced annually. At this rate a 25 kW-output Stirling power conversion unit would be expected to cost about $2,500, meaning the total cost of an installed 25 kW solar Stirling unit would be about $25,000. What is not yet known is how much the cost per unit would decrease if 5 or 10 million solar genset systems were produced annually.

It is worth noting a typical automobile engine that has a wholesale cost of around $1,000 generates about 135 horsepower. As such, it has an electrical potential of about 100 kW, which means the installed cost per kW is only about $10. This, in turn, means a 25 kW output engine would be expected to cost about $250, rather than the $2,500 cost that is assumed if only 25,000 Stirling engines are produced annually. Although the cost of the concentrator itself may not be reduced by a factor of ten, it is reasonable to expect that the total cost of a 25 kW solar engine system (that weighs about 15,000 lbs.) could be reduced to about $15,000 (i.e., $600/kW) once it is mass-produced.

Because solar engine systems do not require the purchase of chemical fuels, this cost is competitive with the capital costs for centralized fossil fuel or nuclear facilities, which can range from $500 to $1,000 per installed kW. Solar engine systems will have no fuel costs, and therefore no fuel cost increases. As a result of all of these factors, when solar Stirling genset systems are optimized, they should be able to generate electricity for less than 4 cents/kWh. This cost is competitive with the electricity generated from fossil fuel or nuclear facilities, especially given that solar units will provide the electricity during times of day when the demand is the highest.

BioStirling Systems

Biomass energy resources are made up of any organic matter available on a renewable basis for conversion to energy. Converting biomass to fuel for Stirling engines can be accomplished by direct combustion of the dry biomass feedstocks, or biochemically converting the wet feedstock materials via anaerobic digestion to biofuels such as methane and/or hydrogen. Biomass resources are substantial. According to data published by the U.S. Department of Energy's National Renewable Energy Laboratory (NREL), U.S. annual energy demand is roughly 95 quadrillion Btus (Quads), and annual world demand for energy is currently about 350 Quads. NREL has calculated the current global biomass resource to be in excess of 2,700 Quads. Thus biomass resources could totally displace fossil fuel and nuclear energy systems. In the U.S., both SES and STM are commercializing "BioStirling" genset systems that use wood chips or other biomass resources as fuel. The SES/Boeing team, which is also developing utility-scale renewable energy power plants, is planning to integrate BioStirling systems with other renewable energy technologies in order to provide "dispatchable" power.

Figure 6.31: Biomass Burner
Woodchips or other biomass fuels are placed into the hopper and then fed into the biomass burner system. The image is provided courtesy of Carbon Cycle Corporation, Davis, California.

Thermoacoustic Stirling Engines

One of most fascinating Stirling engines is being developed by investigators at Los Alamos National Laboratory. It is referred to as a "thermoacoustic" engine because it converts sound waves into electricity. This concept was initially conceived in 1979 by Peter Ceperley, a physics professor at George Mason University in Fairfax, Virginia. In the May 2000 issue of *Discover* magazine, author Brad Lemley provided a fascinating overview of Ceperley's work. In his article, Lemley explained that Ceperley was able to reason that the work done by heat in a Stirling engine to expand a gas could also be carried out by sound waves. Humans hear sound waves because the pressure they exert vibrates the eardrums at varying frequencies. Ceperley realized that sound waves could also impact gas molecules and cause them to move back and forth in a classic Stirling-cycle.

In 1997, Scott Backhaus, a postdoctoral student at Los Alamos, was assigned to construct a device to actually test Ceperley's thermoacoustic theory. Backhaus proceeded to build a test engine that he referred to as a Thermoacoustic Stirling Hybrid Engine (TASHE), starting with a baseball-bat shaped resonator made from inexpensive steel pipe. The TASHE power conversion unit performs the same basic task as any ordinary car engine, which is to convert heat into motion. However, the TASHE system operates entirely on pressure waves, using high-intensity sound to do the work. As a result, it has no moving parts; it can be manufactured from relatively inexpensive and readily available materials; and the TASHE system is just as efficient as the conventional internal combustion engines. The resonator determines the operational frequency of the engine in the same way that the length of an organ's-pipe determines its pitch. At the handle end of the bat, Backhaus bolted on a doughnut-shaped metal chamber to hold the hot (700 degree Celsius) and cold (70 degrees Celsius) hydrogen or helium gas that is contained within the heat exchangers. The heat exchangers act like stereo speakers. They create sound, which is then sent down the resonator where it is repeatedly amplified until the gas pressure wave has enough force to power the engine. While one would expect such an engine to be very noisy, the sound waves are mostly absorbed by the quarter-inch thick steel walls that are required to contain the highly compressed working gas.

Thermoacoustic sound engines in a home would be able to use the waste heat from gas fired hot water heaters to generate electricity. In addition, the Los Alamos thermoacoustic investigators are working in cooperation with Cryenco Inc., a large natural gas transport firm located in Denver, Colorado, to build and operate a 40-foot tall commercial-scale thermoacoustic engine that will be able to liquefy up to 500 gallons of natural gas per day. The heat energy needed to run the engine will come from some of the natural gas that is being liquefied. Indeed, one of the remarkable aspects of virtually all Stirling-cycle engines is that when they are operated in reverse, (i.e., electricity is used to operate the engine in reverse) they work as a refrigerator. These Stirling "cryocoolers" are able to liquefy low-temperature cryogenic gases such as nitrogen or hydrogen, and such units are small enough to be placed into a family garage as part of an "in home" hydrogen filling station. The Stirling Cryocooler pictured in Figure 6.32 is a single piston Stirling engine that is powered by an electric motor.

Figure 6.32: Stirling Cryocooler
Stirling engines have the remarkable ability to be operated in reverse to liquefy gases such as nitrogen or hydrogen. The photograph was provided courtesy of Stirling SCR, the Netherlands.

Central Receiver Systems

Central receiver systems, which are also referred to as "power towers," integrate large numbers of flat mirrors, called "heliostats," (refer to Figure 6.33) that are individually focused onto a central receiver tower (pictured in Figure 6.34). Central receivers were initially thought to be the most practical for utility-scale applications because they utilize a centralized power conversion system that could be integrated with a conventional fossil fuel power plant. Such a centralized system also allows a salt storage system to be integrated to allow the power conversion unit to operate for several hours after the Sun has gone down. Central receiver systems, however, have three significant disadvantages that the smaller solar engine systems do not have:

1. Central receiver systems suffer from cosine losses, which simply means that due to the angle of refraction in the heliostats, the power output disproportionately peaks at noon, and is substantially less in the early morning and toward the end of the day when the Sun is at a steep angle relative to the large array of heliostats. Solar engine gensets do not have this problem because their power conversion units are always held directly perpendicular to the Sun.

2. Central receiver systems can only be constructed in relatively flat areas so the large array of heliostats can be accurately focused on the receiver tower. The smaller and modularized solar engine systems, on the other hand, can be installed and efficiently operated on hilly or uneven surfaces.

3. Because of the centralized PCU, if the central receiver plant goes off-line for any reason, the entire electrical output of the plant is lost. By contrast, if a modularized PV or solar engine power plant has a problem with one of the individual PCUs, the remaining units continue to operate independently. Moreover, during the construction phase of the power plant, modularized systems can be placed on-line as they are installed, thereby providing revenue while the plant is being completed.

Figure 6.33: A Solar Heliostat
An individual heliostat designed by General Motors.

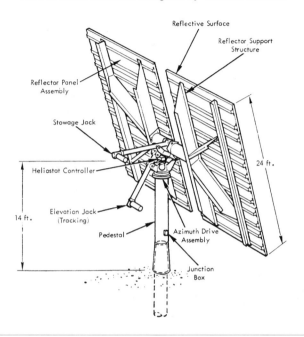

Figure 6.34: Central Receiver System
Multiple heliostats create a large point-focus concentrator system that is focused on the central receiver located in the center. The photograph was provided courtesy of the Southern California Edison Company.

Line-Focus Systems

At present, roughly 90 percent of the electricity generated by solar energy does not come from photovoltaics or solar engine systems, but line-focus systems. Roughly 300 MW of these systems were installed in the 1980s in California by Luz International Limited, a now defunct solar engineering firm that was headquartered in Los Angeles, California. Unlike point-focus concentrator systems that are used with solar engine and central receiver systems, which focus the Sun's energy into a small circular aperture area, line-focus reflector assemblies focus the solar radiation onto specially coated steel pipes, which are mounted inside vacuum-insulated glass tubes. Inside the heat pipe, that runs the length of the trough-like solar concentrator, is a heat-transfer fluid, typically oil, that is heated to about 735 degrees Fahrenheit. The hot oil is then circulated through a heat exchanger, which then heats water until it becomes superheated steam that can then be used to power a conventional steam turbine. While trough systems are only about 11 percent efficient, they have the advantage of being able to be co-fired with conventional hydrocarbon fuels, such as natural gas.

Luz originally had signed a contract to provide electricity at a substantial premium to Southern California Edison (SCE), but when SCE did not renew the contract premium in the late 1980s, the Luz Corporation was forced into bankruptcy. The Luz trough systems were sold to several new operating groups, which have continued to operate the plants. The only major incident occurred in 1998 when an explosion and fire was started by a leak of the hot oil in one of the trough plants. No new trough plants have been built in the U.S. since the 1980s. Solar trough systems have demonstrated that they can work with an availability factor of over 95 percent. However, the capital costs of the solar trough systems are around $3,000 per installed kW, which is five to six times more expensive than fossil-fueled power plants. With ongoing improvements, the capital costs for solar trough systems are expected to be reduced to $2,500 per kW, but in order to make hydrogen competitive with gasoline, the capital costs would have to be reduced to less than $1,000 per kW, and at present, there is only one renewable energy technology that has demonstrated the potential to generate electricity at these low costs: wind energy conversion systems.

Figures 6.35 & 6.36: Solar Trough Systems
Over 300 MW of solar trough concentrator systems have been installed
in Southern California. The photographs were provided courtesy of
Luz International Limited.

Comparative Analysis

One of the more interesting multi-year field tests of the major solar energy technologies was conducted by Southern California Edison in the 1980s at its solar test site located near Barstow, California. The principal investigator at Edison was Charles Lopez, who was responsible for the installation and operation of the four major types of solar technologies: Solar Trough, Central Receiver, Solar Stirling and a double axis tracking Photovoltaic cell array that incorporated a Fresnel lens concentrating system. All four systems operated side-by-side for over three years, which afforded Lopez and Ken Stone, the solar Stirling program manager for McDonnell Douglas (now Boeing), the opportunity to undertake a comparative analysis of the four systems. Their data, which is summarized in Figure 6.37, was correlated on the basis of kilowatt hours generated per square meter per year, to normalize the fact that the respective systems each had a different power rating (i.e., the Central Receiver was a 10 megawatt system, whereas the solar Stirling unit was only rated at 25 kilowatts. The results of the study, that were later published by Sandia National Laboratories, showed that the solar Stirling technology had by far the best performance, and the photovoltaic system had the poorest performance. In spite of this utility test data, DOE has provided the most funding for photovoltaics and the least amount of funding for the solar Stirling technologies.

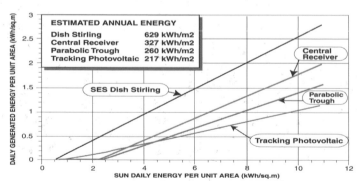

Figure 6.37: Comparative Analysis of Solar Technologies
According to the field data, the solar Stirling unit was generating nearly twice as many kilowatt hours of electricity per square meter per year as the next closest competitor, which was the Central Receiver system.

Wind Energy Conversion Systems

Wind machines were the first solar technology to be used for mechanical power, and they are presently one of the most cost-effective renewable energy technologies. One of the first uses of wind machines occurred around the year 650 AD when Persian millwrights figured out how to use energy from the wind to pump water for irrigation. To this day, similar windmills are still used to pump water. Windmills were first used to pump water in Europe in the early 1100s, and they came into widespread use in the U.S. for water pumping during the early 1800s.

With the development of electrical generating equipment in the late 1800s, both European and American engineers began to experiment with using the wind to operate electrical generating equipment. It was not long before windmills evolved into wind generators. One of the first investigators to develop wind-powered electrical generators was a Danish professor, Poul La Cour, who experimented with wind systems from 1891 until his death in 1908. He was one of the first individuals who foresaw the use of hydrogen as a fuel and the use of wind-powered electrical generators to electrolyze hydrogen and oxygen from water. Another early investigator who promoted wind-powered hydrogen production systems was J. B. S. Haldane, a British biochemist who predicted in 1923 that England's future energy problems would ultimately be solved by constructing large numbers of metallic wind generators in offshore applications that would be able to supply high-voltage electricity to large electrical mains for hydrogen production. During World War II, Vannovar Bush, a physicist who was the director of the U.S. Wartime Office of Scientific Research and Development, became worried over dwindling American fuel reserves and concluded that wind generators might be an answer. He appointed Percy H. Thomas, a wind power advocate, to the Federal Power Commission, which later convinced the Department of the Interior to construct a large prototype wind machine. However, in 1951, the idea died in the House Committee on Interior and Insular Affairs. Wind-generated electricity could not compete economically with coal that was selling for $2.50 per ton or diesel fuel that was $0.10 per gallon. The promise of even less expensive electricity ("too cheap to meter") from nuclear power plants resulted in the abandonment of virtually all federal programs to develop wind-powered energy systems.

In retrospect, this decision turned out to be an unfortunate mistake. Although nuclear plants were initially predicted to cost between $250 and $300 per installed kilowatt (kW), in contrast to wind energy conversion systems that were expected to cost $400 to $500 per installed kW, the nuclear plants ended up costing $1,500 to $2,500 per installed kW. Moreover, the capital costs for nuclear power plants did not include the extensive "front-end" costs of uranium enrichment, nor the "back-end" costs associated with decommissioning the plant at the end of its useful life, or the long-term waste storage problem. In contrast, a NASA study completed in 1972 found that wind machines are a historically mature technology that is based on the straightforward application of well-understood principles and practices in areas of civil, electrical, structural, corrosion and aerodynamic engineering. As a result, any manufacturing problems likely to arise can be subjected to a body of existing engineering knowledge and experience, and thereby rapidly resolved.

The validity of this statement is supported by the fact that wind energy conversion systems have been the most commercially successful of any of the renewable energy technologies in recent years. According to the February 2000 publication of Power Engineering, over 3,600 MW of wind energy systems were installed in 1999, bringing the total worldwide installed wind capacity to 13,400 MW. The top three markets for wind systems in the world are presently Germany, which now has over 4,000 MW installed, followed by the U.S., which now has over 2,500 MW installed, and Denmark, which has just over 1,700 MW installed. According to the European Wind Energy Association, over 40,000 MW of wind energy systems are expected to be installed by the year 2010 and 100,000 MW is expected to be installed by the year 2020. As a result of a 20-year developmental effort by major wind manufacturers, such as Vestas and AEG Micon in Denmark, and EnronWind Corporation in the U.S., the levelized electricity costs from the current generation of wind machines are in the range of 4 to 5 cents/kWh. Additionally, a 1.5-cent/kWh federal tax credit was renewed by the U.S. Congress in 1999. The resulting 3-cent/kWh-electricity cost is generally competitive with conventional fossil fuel power plants operating on natural gas fuel. Moreover, the estimates from the major wind manufacturers are that by 2010, electricity costs will fall to between 2 and 3 cent/kWh with no federal tax assistance.

Figure 6.38: Vestas 660 kW Wind Turbines

Figure 6.39: The Vestas Power Conversion Unit
A Vestas "train-car" size Power Conversion Unit.

The electricity costs from natural gas-fueled power plants, by contrast, will be expected to be continually increasing over their 30-year design life. This is because natural gas is a non-renewable energy resource, and as the supplies are depleted, its cost, and therefore the cost of electricity from natural gas, will increase accordingly. Vestas, which is the largest and most experienced wind manufacturer, has over 2,500 MW of wind machines installed worldwide. The current Vestas wind machines have rated outputs ranging from 660 kW to 1.5 MW. Vestas has over 2,000 employees and production facilities in Denmark, Germany, India, Spain and Italy; and Vestas has plans to have a U.S. manufacturing operation underway by 2002.

Most utilities in the U.S. are now investing in natural gas-fueled combined cycle combustion turbines for base-load electricity production. Permit applications for over 10,000 MW of new fossil-fueled power plants in the Phoenix, Arizona area have been recently filed with the Arizona Corporation Commission, and similar applications have been submitted to the respective utility regulatory commissions in Nevada and California. These new fossil-fueled plants will emit thousands of tons of nitrogen oxides, carbon monoxide, carbon dioxide, volatile organic compounds and particulates annually. Moreover, these emissions will be introduced into the already polluted atmosphere in cities like Phoenix, Los Angeles and Las Vegas. Highway funds are already being threatened because these cities and their surrounding communities are not in compliance with the current EPA air quality requirements. In addition, the 30-year design life fossil fuel power plants will consume roughly 100,000 acre-feet of water annually, and much of that water is intended to come from already depleted groundwater supplies. The problem is compounded because most of the electricity from the new power plants in Arizona and Nevada will be exported outside the state. Thus, the residents in Arizona and Nevada would get the pollution while residents in other states would get the bulk of the electricity. Urban air pollution is already one of the major issues before the Arizona Legislature because of its negative effect on tourism, which is the largest industry in Arizona. Given these already serious air quality problems, if the new fossil-fueled power plants are approved, the air quality -- and the quality of life in general over the next 30-years -- will only deteriorate in metropolitan cities like Phoenix, Los Angeles and Las Vegas.

Given the successful development and deployment of wind energy systems, permits for new fossil-fueled power plants and the related transmission systems should not be granted -- unless it can be shown that the renewable energy technologies and resources are not able to displace such power plants. Moreover, if any new combustion turbine power plants are approved for development, their combustion systems should be optimized to utilize hydrogen fuel in order to minimize any emissions from such power plants. There are substantial wind resources in most countries around the world, and within a number of states in the U.S. According to data compiled by the American Wind Energy Association, the wind potential in the following states is more than enough to provide for all of the projected electricity growth anticipated for the entire U.S. by 2020 (i.e., 400,000 MW):

States	MW Potential
North Dakota	138,400
Texas	136,000
Nebraska	99,100
Wyoming	85,000
Colorado	54,900
New Mexico	49,700
California	6,700
Nevada	5,700
TOTAL	575,000

While states like California have an excellent market for wind-generated electricity, as the above data shows, California's wind resource is relatively limited. This is compounded by the fact that installing wind turbines within the state of California is much more problematic due to siting and aesthetic considerations. This being the case, if large-scale wind power is to be used in California, it is anticipated that the bulk of the wind turbines will need to be installed in remote areas of the surrounding states of Nevada, Colorado, and New Mexico. Once generated, the electricity can then be transmitted to the major urban areas within these states, and the remaining power can then be exported to states with minimal wind land resources, such as Arizona and California.

Although wind power systems have many significant advantages, including a huge international resource that could displace fossil fuel and nuclear energy systems, there are two primary factors that are limiting their deployment. The first issue is related to dispatchability, and the second revolves around a subset of issues that are related to siting.

Dispatchability

The issue of dispatchability has to do with the fact that winds and wind strength are intermittent in nature. As a result, wind turbines may not be operating when they are needed, or conversely, they may be operating in the middle of the night when the electricity is not needed. In addition, the quantity of wind-powered electricity that can be readily integrated into a country or region's electricity grid depends mainly on the system's ability to respond to fluctuations in wind energy supply.

Numerous assessments involving modern utility grids have shown that no technical problems will occur by having up to 20 percent of the grid electricity provided by intermittent wind systems. Studies in Denmark have shown that up to a 50 percent penetration level of a grid system is possible if the electricity can be interchanged with the grid systems in neighboring countries, such as Norway and Sweden, both of which have large capacities in hydropower. As such, Denmark is planning to have 50 percent of its electricity produced by wind systems by the year 2030. However, a more fundamental solution to the intermittent nature of the solar and wind resources involves using all or part of the electricity to manufacture hydrogen from water. The hydrogen can then be stored, transported by pipeline, ship or tanker trucks and later used as a pollution-free fuel for the existing power plants that are presently operating on natural gas fuel.

As the wind and other renewable energy technologies are increasingly mass-produced on a scale to displace new fossil-fueled power plants, their electricity costs will continue to be reduced. This, in turn, will allow hydrogen fuel to be produced at a price that is competitive with hydrocarbon fuels such as gasoline and diesel fuel. This will result in even higher production levels of the renewable energy technologies, which will further reduce the costs of manufacturing electricity and hydrogen.

As Chapter 5 discussed in some detail, hydrogen can be manufactured from water with any source of electricity, and it is an ideal method of storing the vast but intermittent supply of solar and wind energy resources. This is because hydrogen is a universal pollution-free fuel that can be integrated with virtually any energy system, including automobiles, aircraft and power plants. Virtually all green plants on the Earth have been successfully using a solar hydrogen energy process for over 3 billion years, and hydrogen is the only carbon-free combustion fuel that could displace gasoline on a global scale -- forever.

Fortunately, a number of major automotive companies, such as BMW and Ford, are modifying existing internal combustion engines to operate on liquid hydrogen fuel, which most closely resembles gasoline from a standpoint of vehicle range, weight and fuel tank size. Moreover, NASA, which has used liquid hydrogen as rocket fuel since the 1940s, rates hydrogen safer than gasoline in the event of accidents. Whereas 18 gallons of water is required to manufacture a gallon of gasoline from crude oil, only 2.3 gallons is required to make a similar energy content of hydrogen if it is extracted from water with electricity (i.e., a process called electrolysis). However, if hydrogen fuel is to be manufactured from water at a price that is competitive with gasoline, the electricity costs need to be 1 to 2-cents/kWh. With an electricity cost of 2-cents/kWh, the gaseous hydrogen will cost roughly $6 per million Btu (mmBtu), which is similar to the current refinery wholesale cost of gasoline.

The cost of electricity generated from natural gas, by contrast, is expected to only increase over time as the non-renewable reserves of natural gas are exponentially diminished. According to the U.S. Energy Information Administration, the total discoveries of natural gas in 1998 were down 27 percent from 1997, and new discoveries in 1998 were less than half of the new discoveries in 1997, and 30 percent less than the prior 10-year average. U.S. natural gas production in 1998 was 19 trillion cubic feet per well per day, well below the record-high 21.7 trillion cubic feet in 1971. Gas well productivity peaked in 1971, and then fell steeply through the mid-1980s. Although natural gas is an important "bridge-fuel" to the hydrogen economy, the above statistical data on the declining natural gas and crude oil reserves would suggest that the transition to renewable hydrogen technologies should be pursued as soon as possible.

Siting

Issues related to siting include the fact that topographic features on land can dramatically reduce wind speed, which significantly reduces the electrical output of a wind machine. This is because wind-powered machines conform to the "Law of the Cube." This mathematical formula provides that the power content of the wind is proportional to the cube of the wind speed. This means that when wind speed is reduced by 50 percent, the power output of a wind turbine is reduced by a factor of 8. Other issues related to siting are primarily aesthetic in nature. The current generation of MW-scale wind turbines are roughly 250 feet tall and can be seen for many miles. This is compounded by the fact that some of the best locations to place wind systems is highly visible ridge areas where the winds generally intensify.

In order to overcome the problems associated with the siting of wind systems, advanced "Multi-Array" wind systems have been proposed by Ocean Wind Energy Systems (OWES), located in Amherst, Massachusetts, which has been developing the concept since the 1970s. The concept involves placing multiple wind turbines on a single tower which can increase the power output from an existing site by a factor of 10 or 20.

The principal OWES design engineer is William E. Heronemus, who graduated from the U.S. Naval Academy and worked as a naval architect in the largest ship design group in the U.S. until he retired from the Navy in 1965. Heronemus then served as a professor and Associate Head of Civil Engineering at the University of Massachusetts at Amherst (UMA). During his tenure at UMA, Heronemus and his associates developed a Multi-Array wind turbine design that could also be deployed in off-shore applications. Such a system helps to resolve the siting issues in two important ways.

First, a land-based Multi-Array wind system has the potential to increase the power output from an existing wind site by a factor of 10, while having a reduction in capital and balance of plant costs. Moreover, due to the height of a Multi-Array system (500 feet), its overall plant capacity factor will be improved. This is because the wind speed and intensity increases exponentially from the surface of the Earth, and as the wind speed doubles, the electrical power output of a wind machine can be increased by a factor of eight. Second, Multi-Array wind systems lend themselves to offshore operations, which would greatly reduce siting issues. Such offshore systems could provide enough electricity and/or hydrogen to power the entire U.S., and virtually every other country. While the capital costs of sea-based systems are expected to go up by a factor of 1.5 to 2 over land-based systems, these higher capital costs are expected to be offset by the fact that the offshore winds are, on average, much stronger. As a result, the power output of an offshore wind system would be expected to roughly double.

Offshore Wind Systems

Offshore wind systems were first proposed in the 1920s by engineers in the UK, but it was not until the late 1970s that the initial feasibility studies on siting, technology and economics of large-scale offshore wind turbine systems were carried out by investigators in a number of countries, including Germany, Denmark, the Netherlands, Sweden, the UK, and the U.S. With the successful development of land-based wind systems in the 1980s, and the increasing awareness that good land locations (particularly in Europe) were limited, the interest in offshore projects intensified.

Figure 6.40: Existing Offshore Wind Systems
Note that in existing offshore wind installations, the tower that supports the PCU is firmly placed into the seabed floor.

Multi-Array Wind Systems

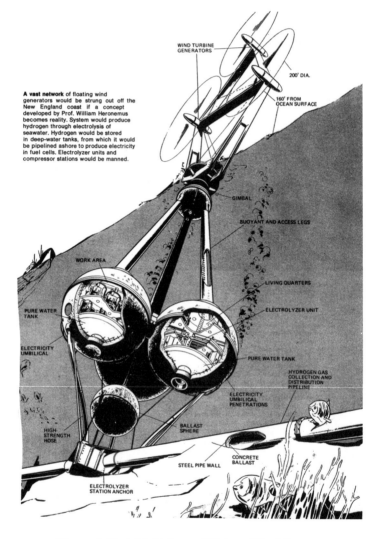

Figure 6.41: A Multi-Array Offshore Wind System
Designed by William E. Heronemus, Department of Engineering,
University of Massachusetts at Amherst. Reprinted from *Popular
Science* with permission © 1977 Times Mirror Magazines, Inc.

Figure 6.42: 36-Array Offshore System
The illustration was reprinted with permission from the *National Geographic Society*, which published the image in December of 1975.

In the 1990s, over 30 MW of offshore wind systems were installed off of the coasts of Denmark, Sweden and The Netherlands. However, the Multi-Array offshore "windship" system that was developed by Heronemus is a floating concept, rather than the existing approach of installing wind turbines on the seabed floor as shown in Figure 6.40. The wind-generated electricity would be fed to undersea hulls where electrolyzer units would be located to produce hydrogen, which would either be picked up by cryogenic tankers, or a pipeline system would carry the hydrogen fuel to shore. A Multi-Array system would typically consist of a large network of wind turbines that would be attached to floating platforms as pictured in Figures 6.41 and 6.42. These units would be arranged in concentric rings around a centralized electrolyzer station that would be used for hydrogen production. The hydrogen that would be produced from the seawater would then be piped ashore and used as fuel for transportation vehicles or the production of electricity in conventional power plants. Each 36-wheel array would be able to generate from 10 to 18 MW, thus roughly one million of these "windships" would be able to generate enough electricity and/or hydrogen to make the U.S. energy independent and essentially pollution-free.

Although offshore wind energy systems would involve some disturbance of the seabed during construction, no disturbances would occur during normal operations. Offshore wind systems could be a significant factor in protecting the ocean's marine organisms that exist on the continental shelves. The fisheries in the oceans are virtually being eliminated by the over-fishing that is made easy by low technology "driftnetting" practices whereby a ship will deploy fishing nets that are miles long, as well as state-of-the-art electronic fishing vessels that utilize radar to track and acquire fish. A generation ago, New England fishermen were taking in thousands of tons of haddock and cod every year, and they were hardly denting the supply. But today, the haddock is gone from the New England coast, just as the herring has been hunted out of the vast North Sea. In what has been referred to as "strip-mining the seas," thousands of fishing vessels are now devastating the once fertile fishing areas on a global scale. More and more of these ships are being put into service every year, but if large numbers of wind turbines were deployed at sea, they would provide an important fish sanctuary that would help to offset this exponential destruction of the Earth's ocean ecosystems.

Vertical Vortex Wind Systems

Another advanced wind energy system that resolves the basic problems associated with traditional land-based systems has been developed by Dr. James T. Yen when he was working as a fluid dynamics research engineer at Grumman Aerospace Corporation. Yen called his innovative system a "vertical vortex generator," although it has been referred to as a "tornado turbine" because it operates on a similar principle. Yen's system operates by constructing a large tower, and as wind enters the tower through louvers in its side, a vortex of air is created. Because of the low pressure of the vortex, air is sucked in through the openings in the bottom of the tower, which then spins a turbine to generate electricity.

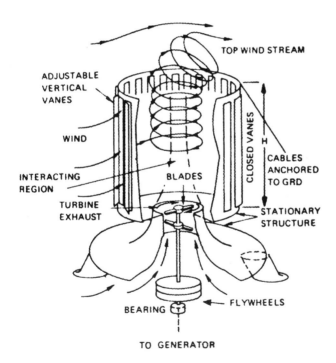

Figure 6.43: Vertical Vortex Wind System
The image was reprinted from *Popular Science* with permission
© 1977 Times Mirror Magazines, Inc.

The artist's conception in Figure 6.44 provides a view of what a vortex energy system might look like in a major metropolitan area. By placing vortex generators in urban areas, they can take advantage of the higher than normal winds that are often generated by large skyscrapers, which then channel the wind into the narrow passageways.

Figure 6.44: An Urban Vortex Generator
Reprinted from *Popular Science* with permission © 1977
Times Mirror Magazines, Inc.

A vortex is a swirling mass of air (or water) that forms a low-pressure vacuum at its center, and anything that is caught in the motion is drawn in. This is how whirlpools and tornadoes operate, and the reason they develop such awesome power is because the wind pressure energy in a 15 mph wind is about 3,000 times greater than the kinetic energy that drives conventional wind energy systems. Thus, while a 200-foot diameter blade on a conventional wind turbine can produce approximately one megawatt, the same size blade on a vortex generator could produce from 100 to 1000 megawatts. A vortex generator system is also able to operate if there is insufficient wind by heating air with solar energy or by hydrogen fuel. Although no large prototype vortex generators have been built, they offer a viable option that could be used in major metropolitan areas where conventional wind generators would be impractical. In this regard, architect Paolo Soleri has designed a megastructure city pictured in Figure 6.46 that could have a vortex generator integrated into the heart of the structure.

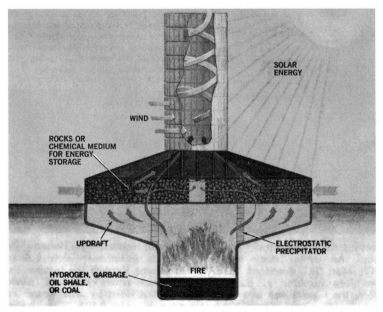

Figure 6.45: Vortex Cutaway
Vertical vortex generators can be powered by a wide-range of alternative fuels, including hydrogen. Reprinted from *Popular Science* with permission © 1977 by Times Mirror Magazines, Inc.

Figure 6.46: A Vortex Arcology.
This megastructure, which is a city without automobiles, was designed by architect Paolo Soleri. The illustration was reprinted with permission from the Massachusetts Institute of Technology.

The idea that wind energy systems could, and should, provide a major portion of the energy demanded by a modern industrial society has been well documented. Many competent engineers in many different countries have shown that the wind resources are vast, and the technical capability to utilize the resource is quite practical. As professor Heronemus has written:

> *"In the oceanic winds, we have a huge energy resource that modern technology can harness to serve our needs on demand. It could be put to use in the very near term, economically, in an aesthetically satisfactory way, and with no pollution of any kind."*

OTEC

Roughly forty-five percent of all solar radiation that reaches the surface of the Earth is absorbed by the surface water of the tropical oceans. Thus, the oceans are the largest solar collector in the world. Although the heat capacity of water is greater than that of any other fluid, the energy from the Sun cannot penetrate very deeply into ocean water. This is why cold water underlies even the warmest of the tropical seas. Ocean Thermal Energy Conversion (OTEC) systems not only can tap this enormous energy potential, but they are the only solar technology that can operate 24-hours a day, seven days a week, regardless of the time of year or weather conditions. This is because OTEC systems operate by taking advantage of the constant temperature differential that exists between the solar-heated surface water (which is about 80 degrees Fahrenheit) and the cold deep water (which is about 40 degrees Fahrenheit) to produce energy.

The Second Law of Thermodynamics states that the conversion of heat into mechanical work is possible when two heat reservoirs of different temperatures are at one's disposal. Hence the tropical oceans become a prime choice for thermal energy conversion systems. They have enormous heat capacity and a constant temperature differential of about 40 degrees that is permanently maintained by natural forces. OTEC systems, which are similar to large ships, are relatively simple to manufacture and maintain because they operate in relatively low-temperature seawater environments. This means that any company that can build a supertanker or an offshore drilling rig can build an OTEC system. Capital costs for OTEC plants have been estimated to range from $1,000 to $3,000 per installed kW, resulting in electricity costs from 2 to 8 cents per kWh. The variation in these capital costs is a reflection of different designs and number of units that would be produced annually. The mini-OTEC plant pictured in Figures 6.50 and 6.51 was developed by Lockheed under contract with the U.S. Department of Energy during the Carter Administration. Although the tests proved successful, the program was terminated when the Reagan Administration concluded that the free market forces and not the federal government should determine national energy policy. The problem with that reasoning is that when federal funding for OTEC research and development stopped, so did the research and development.

Figure 6.47: Ocean Thermal Energy Conversion
Courtesy of Lockheed Missiles & Space Company, Inc.

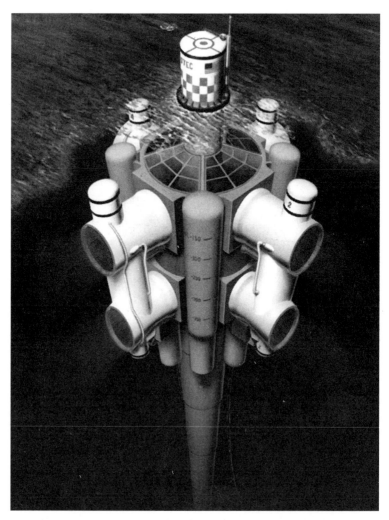

With OTEC systems, the heat from the warm surface water is used to boil a working fluid, which creates a high-pressure vapor that is then used to spin a turbine. The turbine then drives a generator to produce electricity and/or hydrogen. The vapor is then chilled by the cold sea water that lies roughly 600 meters (2,000 feet) below the surface, until it turns back into a liquid, at which time it is then pumped back to repeat the cycle.

Figure 6.48: Lockheed OTEC Cutaway
Courtesy of Lockheed Missiles & Space Company, Inc.

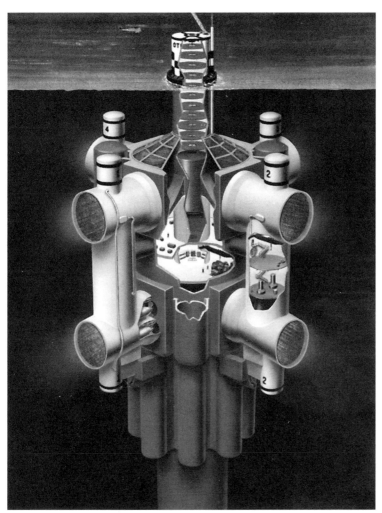

The same basic system is used in conventional coal or nuclear power plants; the only difference is the working fluid. Where conventional power plants use water as a working fluid to make high-temperature steam, OTEC designs use a liquid that boils at or below the seawater's surface temperature, such as ammonia or propane. Since OTEC systems use a closed cycle, none of the working fluid ever escapes during normal operation.

Figure 6.49: OTEC Operation
Courtesy of Lockheed Missiles & Space Company, Inc.

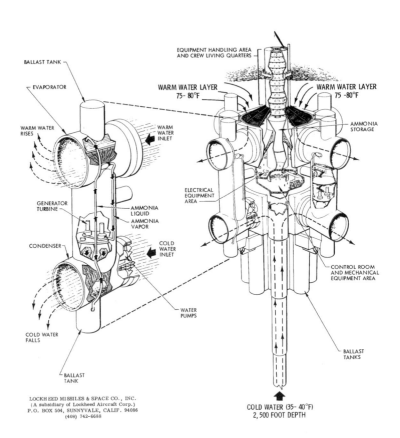

Note that in Figure 6.49 the near freezing seawater that is pumped up into the power conversion units is discharged well below the warm surface waters. Likewise, the warm seawater near the surface is pumped through the warm water heat exchangers that are also located near the surface. The Mini-OTEC pilot plant developed for the U.S. Department of Energy by Lockheed Missiles & Space is pictured in Figures 6.50 and 6.51. Although the Mini-OTEC unit was successfully demonstrated off the coast of Hawaii in the 1980s, the U.S. Congress and DOE eliminated any further funding for OTEC in the 1990s.

Figure 6.50: A Prototype Mini OTEC Plant
Courtesy of Lockheed Missiles & Space Company, Inc.

Figure 6.51: Mini OTEC Primary Components
Courtesy of Lockheed Missiles & Space Company, Inc.

Figure 6.52: Grumman OTEC Design
Note the tugboat that is towing out a heat exchanger component of the
OTEC plant. The image was provided courtesy of Grumman
Aerospace Corporation.

In the artist's drawing (Figure 6.52) of an OTEC plant
designed by the Grumman Aerospace Corporation in the 1980s,
note that one of the OTEC plant's modular heat exchanger mod-
ules is being towed to a site by a tugboat. Whereas the OTEC
designs by Lockheed, Grumman and TRW all assume the OTEC
plant will be tethered to the seafloor, the Johns Hopkins
University design pictured in Figure 6.53 actually moves freely
through the water like a conventional ship.

Figure 6.53: Johns Hopkins OTEC Design
The image provided courtesy of the Applied Physics Laboratory,
Johns Hopkins University.

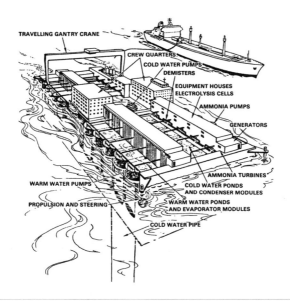

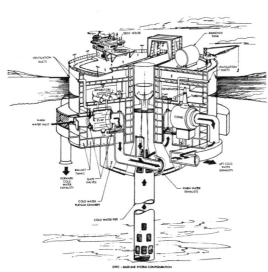

Figure 6.54: TRW OTEC Design
Image provided courtesy of TRW.

Some of the most significant advantages of OTEC systems include the following:

1. OTEC systems, in and of themselves, could produce enough energy, in the form of electricity and/or hydrogen to run the world.

2. OTEC plants (or ships) can be built in large numbers in existing shipyards in a manner similar to the mass construction of the Liberty ships of World War II.

3. Although OTEC systems can be based on land, the vast majority of OTEC systems will be deployed at sea, and as a result, they will not require the purchase of real estate.

4. No high temperatures are involved in OTEC systems. Thus, less expensive materials can be used in their construction.

5. OTEC systems have been designed by numerous engineering firms, including Grumman Aerospace, Lockheed, Bechtel, and TRW. Academic OTEC research teams exist at Carnegie-Mellon University, Johns Hopkins University and the University of Massachusetts.

The OTEC concept is over 100-years old. OTEC was first suggested as a source of power by the French physicist d'Arsonval in 1881, and the first OTEC pilot plant was constructed in 1930 in Cuba. Although the OTEC pilot plant was able to generate about 22 kilowatts of power, the low cost of fossil fuels prevented the project from being continued. There is no question, however, regarding OTEC's vast energy potential, and as W. H. Avery, who is the director of the Ocean Energy Programs at the Applied Physics Laboratory at Johns Hopkins University has pointed out, OTEC plants are not dependent on any high technology breakthroughs. Rather, they essentially involve elaborate plumbing and relatively low-temperature pumping systems. In addition to their obvious purpose of electric power generation, OTEC power plants have two other remarkable advantages.

First, OTEC plants can be designed to desalinate seawater at a fraction of the normal cost. Investigators have calculated that the cost of desalinated seawater from OTEC systems could be as low as five cents per thousand gallons. Thus, it would be less expensive for most major coastal cities to have fresh water barged in from OTEC plants rather than acquire it from their own municipal water systems. The second remarkable characteristic of OTEC power plants is they can greatly enhance fish yields. This is because the cold deep water that OTEC plants pump up to condense their working fluid back to a liquid, is rich in the nutrients that are necessary for aquatic plant and animal life. Indeed, natural cold-water upwellings are responsible for some of the most fertile fishing grounds in the world, such as those off of the west coast of South America. This means that deploying large numbers of OTEC plants throughout the tropical seas could dramatically increase world seafood supplies with vast open sea plant and fish farming areas.

According to researchers at Columbia University's Lamont-De-Herty Geological Observatory, the condenser effluent from a relatively small 100-megawatt OTEC plant could yield about 129,000 tons (wet weight) of shellfish meat annually, and a similar quantity of carrageen-containing seaweed. The current wholesale price of shellfish meat exceeds $2.00 per pound. This being the case, the value of the shellfish alone could exceed $500 million, which is over 5 times the estimated $100 million capital cost of the OTEC plant itself. Given all of the apparent advantages of OTEC systems, one can only wonder why it is nuclear technology that continues to receive the vast majority of energy research and development dollars. One thing is clear: between the winds in the atmosphere and the thermal differences that lie within the oceans, there is a vast resource of pollution-free energy. Heronemus and his Amherst, Massachusetts engineering design teams have worked extensively on both OTEC and wind energy conversion systems, and he summarizes his views on these renewable energy technologies as follows:

> *"These processes are the way of the future. When combined with other solar energy processes, they constitute the only energy regime which can sustain any real growth without making our planet uninhabitable."*

Economic Considerations

If a "fair accounting" system were used that factored in the environmental, health and military costs of burning fossil and nuclear fuels, solar hydrogen energy systems would be the least expensive in a free market system. However, even with the existing accounting system that ignores these "external costs," the renewable energy technologies can compete if they are mass-produced. The miracle of mass production revolves around the fact that in most cases, if more individual units of something are produced, the cost of each unit will decrease accordingly. It is important to realize that when most energy analysts calculate the relative energy costs that would be generated from renewable energy technologies, they make assumptions about how many units will be produced per year. In most cases, there is a general assumption made that renewable energy technologies would provide only a small percentage of electricity production, and electricity only accounts for about 20 percent of the overall energy use of most industrialized countries. As a result, the projected production rates for renewable energy systems were always assumed to be relatively small.

If the renewable energy technologies are mass-produced on a scale to provide sufficient electricity to produce enough hydrogen to displace the use of fossil fuels, their costs will continue to be reduced. Mass-production on such a scale will mean totally committed factories would be designed and built to mass-produce the renewable energy equipment. Moreover, as more and more engineers are able to refine the technology over a period of years, that will also contribute to reduced costs.

Renewable energy technologies are also modular in design. This is important for the following reasons:

1. As new technology is developed to further reduce the cost of producing hydrogen, it can easily be assimilated into a modularized energy system, which inherently lends itself to an era of rapid technological change. This is in contrast to nuclear plants, which can take over 10 years to build. This means that from the time nuclear plants are initially designed to the time they are finally completed, it is likely that they will become both economically and technologically obsolete.

2. Energy systems that are modularized lend themselves to mass-production; they can be purchased incrementally and will have very short construction lead-times. This is in contrast to nuclear plants that require billions of dollars to be financed for the 10 or more years it takes to build the facility. This is an important financial consideration, given the direct relationship of time and money.

3. Modularized energy systems are inherently more reliable than single-source centralized facilities. If one elephant could do the same job as a million ants, which would be better to use for a task? Consider that if the elephant dies or gets sick, the work cannot be accomplished. But if one has a million ants working and several die or are otherwise unable to work, the overall job still gets done. Simply put: There is safety in numbers.

The relative risk factors of the renewable energy technologies are small in comparison to nuclear systems that have no financial bottom line; or the ecological disasters that can result from oil spills and strip mining. While some people might automatically assume that the major oil companies would seek to prevent the use of renewable hydrogen energy systems, in fact, it has been the major oil companies that have helped to sponsor many of the technical conferences that are put together every two years by the International Association for Hydrogen Energy. This should not be surprising.

Senior executives of the major oil companies have repeatedly indicated in formal testimony before the U.S. Congress that they and their stockholders are always looking for legitimate long-term alternative energy investments, provided that the alternative energy systems can provide them with an acceptable renewable rate of return on their invested capital. Executives of the major oil companies know all too well about the difficulty of finding new oil. This explains why they are now finding it is less expensive to simply buy their competitors' oil reserves rather than to find what is left. Thus, rather than oppose a transition to renewable hydrogen production technologies, the oil industry will likely be at the heart of their mass-production when the appropriate national energy policy has been established.

U.S. Energy Policy

Traditionally, the energy policy of the U.S. government has revolved around encouraging private energy companies to explore and extract fossil fuels and other natural resources. As a result of the Arab oil embargo that occurred in 1973, the Nixon Administration established a goal of making the U.S. energy-independent from foreign energy suppliers without specifying how the objective was to be accomplished. The Carter Administration called for the "moral equivalent of war" to be declared in order to make the U.S. less dependent on energy imports. But although the research into solar technologies was substantially increased during the Carter Administration, the bulk of the research dollars involved developing other fossil fuel resources, principally oil shale and coal.

With the election of Ronald Reagan, U.S. energy policy shifted back to finding and consuming what is left of the fossil fuels and building additional nuclear power plants. President Reagan abandoned the national goal of energy independence established by the Nixon Administration. Thus, what little engineering and financial resources that were being focused on the renewable resource problem were shifted to an already substantial arms race, which included a new range of highly complex and expensive "Star Wars" space weapon systems. The members of Congress and the advisors to President Reagan apparently did not realize that their energy and environmental policies were an even greater long-term threat to the security of the U.S. than the Soviet armed forces.

This is because an armed conflict with the Soviet Union was an unlikely event, whereas the problems of resource depletion and environmental contamination were and are very much in the process of occurring. Given the exponentially deteriorating global environmental problems, and the long lead-times and substantial capital investments that are required for making major industrial changes, it is unfortunate that there was not an effort to shift the financial and engineering resources from the arms race to the renewable energy resource race that would directly benefit all countries. Although the Soviet Union disintegrated in the 1980s, the Reagan, Bush and Clinton Administrations continued with an energy policy that seemed to assume that oil and the other fossil fuels were renewable.

U.S. oil imports increased from 30 percent during the Arab oil embargo in 1973 to over 55 percent in1999. Indeed it was not until the year 2000 that oil prices returned to the $30 per barrel range. It is worth noting that Russia and China need to make an industrial transition to renewable resources as much as the U.S. or the European community does. Russia has been one of the largest oil producers in the world for many years, but its oil production peaked in 1984 and is expected to continue declining. U.S. oil production peaked in 1970, and it is also expected to continue declining. As a result, it is in the national interest of all countries to undertake a transition to renewable resources before the diminishing fossil reserves eventually trigger an armed conflict over the remaining Middle East oil reserve. From a U.S. perspective, the possibility of a war in the Middle East is an especially serious concern because the U.S. is not in a strong strategic position in that part of the world. If U.S. military forces could not successfully neutralize Russian or Chinese military forces without the use of nuclear weapons, the situation has the potential to rapidly escalate into the full-scale nuclear exchange that has been dreaded for so long by so many.

It is significant that utility-oriented engineers and planners are generally only interested in electricity production and not normally involved in hydrogen energy engineering research. Very few of the solar and wind engineering teams are aware that the International Association for Hydrogen Energy even exists. Moreover, the bulk of the hydrogen engineering community has been essentially unaware of the most promising renewable energy technologies that can be mass-produced for large-scale hydrogen production. This is a classic case of a significant information-gap between highly trained specialists who must, by necessity, focus their efforts in relatively narrow fields. The problem is compounded by the fact that there are few comprehensive energy research specialists who take a look at the big picture with a long-term view.

The renewable resource investments are going to come from the private sector, and not deficit-financed taxpayer dollars. This is because solar hydrogen and other renewable resource technologies will provide a relatively low risk, renewable rate of return. Both federal and state governments have an important role to play in terms of establishing a national energy policy that will be oriented around a transition to renewable resources.

Although direct costs to state and federal governments will be minimal, a fair accounting system that factors in external costs of energy is needed to provide a fair competitive environment for the renewable energy technologies. Like the war effort in World War II, a cooperative effort between industry and government is going to be required if such a fundamental energy and industrial transition is to occur. The exponential depletion of the existing oil reserves underscores the need to begin quickly and to act with a sense of urgency. Long before the known oil reserves are exhausted, the price per barrel will increase substantially. Whether or not the U.S. is going to have long-term economic stability will depend upon how quickly it can mobilize its national resources -- which are substantial -- to change course.

A national energy policy needs to be adopted that will allow the U.S. to shift from the "Oil Economy" to the "Hydrogen Economy" before the bulk of the world's oil reserves are depleted. If wind energy systems, solar Stirling and OTEC systems were developed in the 1950s and 1960s instead of nuclear-fission facilities, the average cost to build such systems would have been substantially less than if the same facilities are designed and built today. This is because oil prices have increased substantially since 1950, and as a result, so has the cost of everything else. All that has been accomplished by waiting is that the price of most everything has increased.

The national energy policy and objectives are important. If the industrial transition to renewable hydrogen resources and technologies had started in the 1950s, enough of the renewable energy technologies could have been built to have made the U.S. energy independent by 1973, when the first Arab oil embargo took place. In addition, all of the energy pipelines, ships and other infrastructure technologies would already be hydrogen-compatible, and much of the environmental devastation involving oil spills, radioactive waste, the production of greenhouse gases and acid rain could have been avoided. Although many U.S. government policy planners and Members of the Congress did take seriously the warnings of M. King Hubbert and the other geophysicists at the U.S. Geological Survey concerning the eventual depletion of U.S. oil reserves, billions of taxpayer dollars were spent on developing nuclear-fission power plants, and billions more were spend on developing fusion systems -- instead of the renewable energy technologies.

One of the primary explanations as to why the high-risk nuclear option was favored over the simpler renewable energy technologies was that nuclear advocates initially believed that nuclear systems would be able to produce electricity "too cheap to meter." In addition, there were many people in the U.S. government who wanted to provide a more positive public relations image for nuclear technologies and to justify the vast expenditures of money and engineering talent that went into the development of nuclear weapons. The result was an "Atoms for Peace" program that was put forth in an attempt to provide a constructive peacetime use for the nuclear-fission technologies. As it turned out, this was a tragic miscalculation because unlike nuclear energy systems, the renewable hydrogen production technologies could have been optimized and mass-produced with the automotive and shipbuilding technology that existed at the time. Thus, billions of dollars were invested in developing the wrong technology.

Conclusions

Renewable energy technologies provide a realistic method of providing the energy necessary to displace oil and other fossil fuels with hydrogen. Although photovoltaic cells are a promising solar technology option, at present, the electricity they generate to produce hydrogen is not competitive with the other renewable energy options, such as wind, solar Stirling engine and OTEC systems. If the most promising renewable energy technologies are mass-produced on a scale to provide hydrogen fuel, and if a fair accounting system that factors in the external costs of energy is used, the renewable hydrogen will be economically competitive with fossil-based fuels.

The renewable energy industry will rival the automobile industry when production is in full swing. The net result of such an industrial transition will be that essentially pollution-free industries will be producing pollution-free renewable-energy machines that could be utilized by the most advanced industrial nations or the most remote villages in Third World countries. From the consumer's perspective, there is no reason to be concerned about which renewable resource technology will ultimately turn out to be the most cost-effective. May the best engineering teams win.

As more and more improvements are made, it is reasonable to assume that the energy costs will be continually reduced, in contrast to the nonrenewable fossil and nuclear fuels, which will become increasingly expensive as their resources are exponentially consumed.

Fundamental problems require fundamental solutions.

If the U.S. establishes a national policy to make an industrial transition to renewable hydrogen resources, other countries will follow suit because it is in their own self-interest to do so. Germany and Japan already have already established long-term hydrogen implementation programs, and the International Association for Hydrogen Energy held its13th World Hydrogen 2000 Technical Conference in Beijing, China.

In the U.S., rising gasoline prices, that are now over $2.00 per gallon in many parts of the U.S., have once again attracted the attention of the media, the U.S. Congress and the general public. Unfortunately, instead of promoting energy independence with renewable energy technologies, most Members of Congress are advocating more "band-aid" solutions that involve leasing what is left of the remaining offshore and wilderness areas to oil and other energy companies so they can extract the last of the remaining fossil fuel resources.

The result of these shortsighted decisions will be to misdirect valuable time and resources, and degrade the few remaining wilderness areas. There is another path. State and federal regulatory commissions can and should only approve permits for electric power plants that will not emit chemical pollutants or require the consumption of limited groundwater resources or nonrenewable energy resources. Such a decision will result in the mass-production of renewable energy systems, and by 2010 or 2020, such technologies could virtually eliminate urban air pollution by generating hydrogen fuel to displace gasoline and diesel fuel in the transportation sector.

This strategy provides a path for existing utility and oil companies to evolve into solar hydrogen companies that can make the U.S. energy independent and essentially pollution-free. Ironically, the transition from an "Oil Economy" into a "Hydrogen Economy" will substantially increase the need for new electrical generation systems. Indeed, if all of the gasoline and diesel fuel is to be displaced by hydrogen, the size of the existing electric service providers will need to be increased by a factor of 2 or 3.

If the respective corporation commissions and other regulatory agencies preferentially certify renewable energy power plants, the resulting shift in investments by ESP's in the U.S. will initiate a transition that will have global implications. It is important that the transition occurs as soon as possible because everything will only get more expensive over time. Such a transition will provide long-term "prosperity without pollution." The renewable energy technologies reviewed in this chapter document the fact that there are many viable alternatives to the existing fossil fuel and nuclear energy systems. Chapter 7 will review the physical resources that will be required if such technologies are to be implemented on a global scale.

Figure 7.1: Sun & Sea
Solar Hydrogen

CHAPTER 7

RENEWABLE ENERGY RESOURCES

In the previous chapter, some of the most technically and economically viable renewable energy technologies that could be mass-produced for large-scale hydrogen production were discussed. In this chapter, the energy potential of these options will be outlined, as well as the land and water resources that would be required for their large-scale implementation. To put the energy potential of the renewable resources into perspective, the following information is useful:

- According to the U.S. Energy Information Administration, world energy production in 1998 was about 382 quadrillion (quads) Btus. Roughly 327 quads (or about 85 percent of the world total) were generated by fossil fuels; 26 quads (or about 7 percent) were generated by hydroelectric dams, and about 24 quads (or about 6 percent) were generated by nuclear power plants.

- Total U.S. energy production in 1998 was about 73 quads, although total U.S. energy consumption in 1998 was 94 quads. 56.5 quads (or about 77 percent) of the U.S. production came from fossil fuels. Nuclear provided 7.2 quads (or about 10 percent), and hydroelectric dams provided 3.3 quads (or about 4.5 percent).

- A megawatt (MW) is equal to 1000 kilowatts, and a kilowatt is equal to 1000 watts. A kilowatt-hour (kWh) is equal to 3,413 Btu. A British Thermal Unit (Btu) is roughly equivalent to the energy contained in a single matchstick, or approximately one-third watt hour.

Solar Resources

The amount of solar energy that is received by the Earth each year has been estimated to be in excess of 5 million quads. Although roughly one-third of this solar input is reflected back into space by the Earth's atmosphere, that still leaves over 3.5 million quads of input energy annually. This is more than 11,000 times the 382 quads of energy that were consumed in 1998 worldwide. Sunlight carries energy that is equivalent to about 1000 watts per square meter of exposed surface. An important question is: *How is this vast resource distributed with respect to solar energy conversion technologies?* Although offshore wind and OTEC systems do not have requirements for large real estate or water acquisitions, that is not true for the predominantly land-based photovoltaic or solar Stirling powered genset systems. While solar energy may fall on most of the globe, it is not uniform in its distribution. Just as fossil fuel resources have been concentrated in certain geographical areas, the same is true for solar energy resources. A quick analysis of Figures 7.2 and 7.3 will confirm this fact. While this will be an unpleasant thought for those who are hoping that the large-scale use of solar technologies will put an end to the era of large international energy companies, the stockholders of the energy companies will be pleased.

Figure 7.2 provides an overview of the high solar insolation areas of the U.S. Note that solar radiation is measured in kilolangleys. A kilolangley is 1,000 langleys, and a langley is one calorie of radiation energy per square centimeter. This is equivalent to .216 kilowatt-hours (kWh) per square foot or 603 million kWh per square mile for every 200 kilolangleys. As one would expect, the desert areas of the American Southwest, which cover vast areas of Arizona, California, Nevada and New Mexico, receive the highest concentrations of solar energy in North America. Most of the 200 million acres that make up these desert regions are uninhabited. Although numerous Indian reservations are located in these desert areas, it is likely that many of the Native Americans would welcome the renewable income that would be generated from installing large forests of renewable energy production systems. Assuming twelve 25-kilowatt solar engine gensets are installed per acre, about forty million acres would be required to install the roughly 470 million gensets and the related subsystems that would be necessary to produce the current annual U.S. requirements of roughly 95 quads of energy.

Care would obviously need to be exercised as to where the vast forests of solar gensets or wind energy conversion systems would be placed. Much of the Sonoran Desert in Arizona, for example, has a relatively fragile ecosystem in contrast to the Mohave Desert in California, the desert areas in Southern Nevada or the salt flats in Utah. From a solar input standpoint, it would be both desirable and theoretically possible to place all 500 million gensets within the southern portion of Nevada or Arizona. However, given the many economic, environmental, political and strategic-military reasons, it is reasonable to assume that after a careful study of potential land areas has been undertaken, the solar genset systems would be distributed throughout various optimal regions of the South and Southwest areas of the United States and northern areas of Mexico.

Figure 7.2: U.S. Solar Resources
The annual mean total hours of solar energy is shown in langley's.
Source: *Atlas of the United States,* published by the
U.S. Department of Commerce, June 1968.

Two Princeton University investigators, Joan Ogden and Robert Williams, calculated how much land would be required to displace fossil fuels in the U.S. with photovoltaic (PV) hydrogen production systems. The PV efficiencies are assumed to be 15 percent and the efficiency of the electrolyzers is assumed to be 84 percent. The land required is shown as a set of concentric circles in New Mexico. Given that solar Stirling energy conversion efficiencies are 30 percent efficient, the land area required would be reduced by roughly 50 percent. As a practical consideration, such large-scale installations would be distributed throughout a number of states in the Sunbelt regions of the U.S.

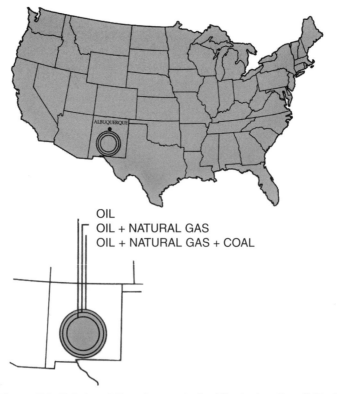

Figure 7.3: U.S. Land Requirements for Displacing Fossil Fuels
Map shows land area needed to replace oil, natural gas and coal in the U.S. with photovoltaic hydrogen production systems. Source: Joan M. Ogden and Robert H. Williams, *Solar Hydrogen: Moving Beyond Fossil Fuels* Published by World Resources Institute, Washington, D.C.,1989.

Figure 7.4: Land Requirements in Saudi Arabia
for Displacing Oil Exports

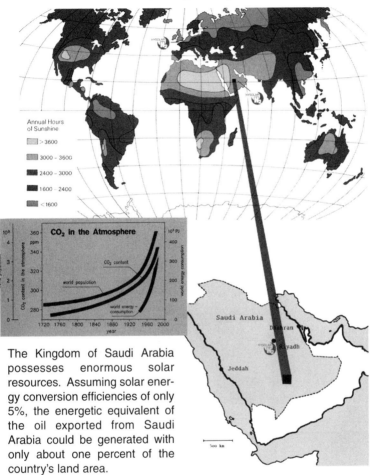

The Kingdom of Saudi Arabia possesses enormous solar resources. Assuming solar energy conversion efficiencies of only 5%, the energetic equivalent of the oil exported from Saudi Arabia could be generated with only about one percent of the country's land area.

While many people assume that oil-rich countries would oppose a transition to solar hydrogen systems, one of the world's first solar hydrogen production systems was installed in Saudi Arabia. The Kingdom of Saudi Arabia understands that while they have the largest oil reserves in the world, those reserves will be substantially depleted by 2050. Given Saudi Arabia's enormous land and solar resources, it is well positioned to take full advantage of the transition to a solar hydrogen energy system.

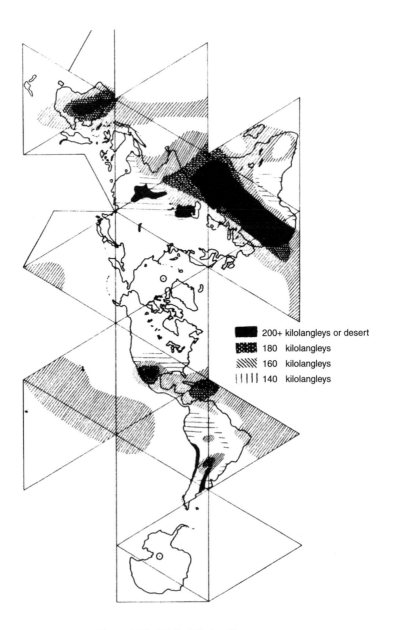

Figure 7.5: Global Solar Resources
Mean Annual Solar Radiation
Source: *Physical Geography: Earth Systems,* 1974.

The fact that solar land resources are concentrated in relatively limited areas of the U.S., and in a larger context, in only limited areas of the planet, ensures that large-scale solar use will allow existing energy companies to maintain their strong economic position. If an energy company is going to be installing solar gensets in large numbers for base-load solar-hydrogen production, it will obviously choose to place the gensets in areas with the highest levels of solar energy to maximize its capital investment. Thus, states like Arizona and California are like Saudi Arabia or Kuwait in terms of their solar energy resource. Unlike oil, however, the Sun's energy will be able to provide energy companies with a renewable rate of return on their investments.

The potential economic implications of large-scale solar development to states like Arizona, California, Nevada, New Mexico and Texas are profound. It is interesting to note that the federal government owns most of the mountains, while the state governments control most of the flat land areas where the renewable energy technologies would likely be placed. As a result, state governments will be able to negotiate long-term lease agreements with energy companies in the same way that grazing rights for animals are negotiated. It is also significant that the renewable energy technologies use a relatively small surface area of the land, which means the land could be co-used for grazing or agriculture. Countries other than the U.S. with substantial land and solar resources include Mexico, Saudi Arabia, Egypt, Israel, China and Australia.

There are a number of heavily industrialized countries that have neither the land availability nor high-intensity solar energy input. These countries include Japan, Germany, France, Great Britain, Sweden and Denmark. However, if hydrogen is produced in areas with high levels of solar insolation, it can then easily be shipped to other countries in much the same way that petroleum and natural gas are presently transported; by pipeline, trains, trucks, and ships. Although the cryogenic tanker manufactured by General Dynamics in Figure 7.7 was initially designed to transport liquid natural gas, essentially the same type of cryogenic vessels would be used to transport liquid hydrogen from production facilities to world markets with great efficiency and without environmental damage. The advanced hydrogen fuel tanker in Figure 7.8 is being currently being developed in Japan by Kawasaki Heavy Industries.

Figure 7.6: A Coastal Liquid Hydrogen Fuel Storage Facility
Although this design was initially planned for storing liquid natural gas
(LNG), similar cryogenic systems will be used for transporting liquid
hydrogen to world markets.

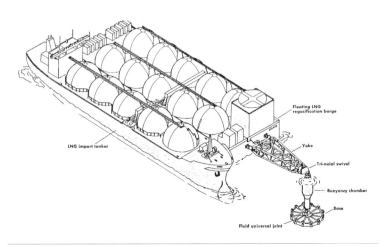

Figure 7.7: A Cryogenic Tanker
With liquid hydrogen transport ships, oil spills from accidents as well as
routine tanker operations will no longer pose environmental problems.
The image was provided courtesy of General Dynamics, Inc.

Figure 7.8: An Advanced Hydrogen Tanker
Kawasaki Heavy Industries currently manufacturers liquid hydrogen
tankers similar to the one pictured in Figure 7.7, but their advanced
hydrogen transport ship design is pictured below.

Figure 7.9: Hydrogen for Japan
The Ministry of International Trade and Industry (MITI), which coordi-
nates Japan's industrial strategy has developed a long-term plan to
convert all of the major ports in Japan to handle liquid hydrogen.

Figure 7.10: Liquid Hydrogen Pipelines

Union Carbide engineers have developed a liquid hydrogen pipeline that could also serve as a superconductor that could also transmit electricity with virtually no energy loss.

Figure 7.11: Liquid Hydrogen Tanker Trucks

In addition to being transported by ship and pipelines, liquid hydrogen can also be delivered from centralized production facilities with fleets of cryogenic trucks like the one pictured above.

As the industrialized countries make a transition to a hydrogen energy system, oil tankers will evolve into cryogenic liquid hydrogen transport ships that will not pollute the oceans during their regular operations. It is not commonly known that most of the oil that is released into the oceans is not due to accidents, but from normal operations where seawater is routinely used to clean excess oil from the tankers. This will not be necessary with liquid hydrogen transport ships. Although the fuel might be lost in the event of an accident, the liquid hydrogen would rapidly vaporize and dissipate harmlessly into the atmosphere. Thus, the ecological damage caused by routine oil tanker operations and accidents would be a thing of the past.

Wind Energy Resources

Wind is defined as the movement of air across the Earth's surface. Winds are primarily the result of the Earth's atmosphere absorbing the immense quantities of energy produced by the Sun. The differential heating of the Earth's surface causes a lateral heat flow that keeps the gases and particles of the global atmosphere in motion.   Other winds are caused by what is referred to as the "evapotranspiration cycle," which involves heat being stored in the atmosphere until it rains, cooling things off, thereby allowing the heat build up to start over again. The principal supply of "fuel" for the Earth's heat engine is in the form of water vapor (which happens to be the primary combustion product of hydrogen and oxygen). Water vapor is a kind of fuel storage system because as it is alternately evaporated and condensed, energy is released to the atmosphere, and the most evident form of this energy is the wind. Although wind energy varies from second to second and is not equally distributed around the globe, for centuries sailors have known that it is remarkably consistent over time within large areas. In general, the major wind systems intensify from equator to pole, with much modification of these patterns due to the relation of land and water masses and topographic features such as mountains, trees or buildings.

The American Wind Energy Association, surveyed the existing data of land-based wind resources in the U.S., and concluded that there was an annual electrical potential of at least 10 billion megawatt hours. This is in contrast to the roughly 3.6 billion megawatt-hours of electrical energy consumed by the entire U.S in 1999. While many people are understandably opposed to placing large numbers of wind machines on highly visible mountain ridges, there are other options discussed in the previous chapter that include placing large numbers of wind systems out at sea. It is also possible to develop the advanced vortex generator configurations that do not need to be placed in mountain top environments. There is no question about the vast potential of wind-powered energy conversion systems. One can only wonder why so many of the industrialized nations have neglected this renewable resource technology for so long.

OTEC Resources

The oceans contain 98 percent of the Earth's water, and they make up over 70 percent of the Earth's surface area that receives solar radiation. This makes the oceans the largest solar collector on the Earth, and it has cost nothing to build. Moreover, half of the Earth's surface lies between the latitudes 20 degrees North and 20 degrees South, which is mostly occupied by the tropical oceans where ocean thermal energy conversion (OTEC) plants could efficiently operate. According to calculations by Clarence Zener, professor of physics at Carnegie-Mellon University, the potential energy that could be extracted by OTEC plants located in the tropical ocean areas would be about 60 million megawatts. William Avery, director of the Applied Physics Laboratory at Johns Hopkins University, has calculated the OTEC potential to be about 10 million megawatts. Assuming the lower value is correct, and assuming the OTEC systems would have an operating capacity of about 80 percent, they would be able to generate about 70 billion megawatt-hours per year. If Zener's more optimistic calculations are correct, the OTEC systems could generate over 400 billion megawatt-hours per year, which is a factor four larger than the current total human energy consumption of roughly 100 billion megawatt-hours. Thus, both wind and OTEC systems could, in and of themselves, generate enough electricity and/or hydrogen literally to run the world -- without using any of the Earth's remaining fossil fuel reserves. It follows that all of the impending environmental problems that will result if those remaining fossil fuels are extracted, shipped and burned could be avoided. Zener has calculated that even if 100 percent of the world's energy needs were provided by OTEC systems by the year 2000, and even assuming the entire world was consuming energy at the rate that the U.S. does, the surface temperature of the tropical oceans would only be lowered by less than one degree Centigrade. For all of these reasons, wind and OTEC energy conversions deserve careful consideration in formulating a national and international energy transition to renewable resources. Once the renewable energy technologies are mass-produced, individual home owners and companies may want to purchase their own systems and become energy-independent. However, for millions of urban dwellers that live in large high-rise cities such as New York or Chicago, they have neither the space nor the appropriate weather conditions for solar or wind systems. As a result, most individuals in these areas will continue to purchase electricity from centralized electric service providers. What may change in the future, however, is that homeowners will be able to manufacture hydrogen in their garage with a small electrolyzer, and then liquefy the hydrogen with a small Stirling cryocooler, and then fill up their hydrogen-fueled car -- all while they are sleeping.

Implementation Lead-Times

There are several key points to consider regarding implementation lead-times that are obvious when one examines Figure 7.12. Note that there are two estimates of world crude oil production. These two curves represent the conservative and liberal estimates of projected world oil reserves. It is also important to note the relative U.S. oil production, which peaked in 1970, is expected to be substantially exhausted by the year 2010. The graph in Figure 7.12 was prepared by Lockheed Aircraft engineers. The purpose of the illustration was to explain to their management as well as congressional and other governmental planners that American industry is spending billions of dollars building airplanes that are designed to burn a fossil-based fuel (aviation kerosene) that will not be economically available for the 20 to 30-year life of an average commercial aircraft. This is a profoundly serious problem that has been essentially ignored by government policy planners.

Figure 7.12: Life-Cycle Considerations
Given current forecasts of oil availability, billions of dollars worth of aircraft are being designed to operate on a fuel that will not be available for the life of the aircraft. The image was provided courtesy of Lockheed Aircraft Corporation.

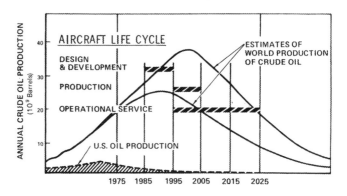

In the 1980s, Lockheed engineers testified to Members of Congress and the Department of Energy that the U.S. should begin designing and building alternative-fueled aircraft that would not be dependent upon fossil fuels. Unfortunately, no action was taken. While Figure 7.12 is concerned only with aircraft, it is important to consider that pipelines, ships, trains, trucks and other long-term capital-intensive industrial infrastructure investments are equally affected by the economic availability of fossil fuels. At present, most of the existing million-mile natural gas pipeline system in the U.S. could accept up to 20 to 25 percent gaseous hydrogen with little or no modifications, but to transport pure hydrogen will involve modifying existing pipelines or laying new ones. This underscores how important it is to have long-range strategic planning because billions of dollars can be saved if the pipelines that are going to be laid in the next 20 years are hydrogen-compatible. If such advanced planning is undertaken, when pure hydrogen is eventually phased in on a large-scale in the future, the enormous investment in the energy pipelines and other infrastructure items will not have been wasted. It should be clear that it is much more cost-effective if a long-term strategic energy plan and timetable is developed and understood by representatives of both industry and government. The failure to plan ahead will only result in crisis and panic reactions, conditions that are hardly conducive to rational thinking and strategic planning.

Water Considerations

One gallon of water has a hydrogen energy content of about 52,400 Btu, compared with a gallon of unleaded gasoline, which has about 120,000 Btu. This means about 2.3 gallons of water will be required to extract enough hydrogen to equal the energy contained in one gallon of gasoline.

"At Gulf, we're working on a way to light lights, cook meals, and heat houses with the energy stored in water."

"You probably remember from grade-school science that water is two parts hydrogen and one part oxygen," says Dr. John Norman.

"Here at General Atomic Company, a subsidiary 50% owned by Gulf Oil, a project is under way to

"*There are 326 million cubic miles of water on earth, and hydrogen in every drop—a natural energy resource that won't run out.*"

extract hydrogen from water for use as a fuel: for heating, cooking, or anything that now uses petroleum or natural gas.

"The extraction process is called thermochemical water-splitting. We know it works because we've done it. But it takes high temperatures — about 1600° F — so it's rather expensive.

"It may be the turn of the century before it becomes commercial. But it's an attractive idea. Hydrogen from a gallon of water has about half as much energy as there is in a gallon of gasoline.

"Hydrogen can be made into a liquid or gaseous fuel. It can be transmitted long distances more cheaply than electricity. And when hydrogen burns, it's converted back into water. Very tidy."

At Gulf, our first priority is to get all the oil and natural gas we can out of resources right here in America. But we're working on a lot of other ideas, too. Thermochemical water-splitting is one of them. We are also working on underground coal gasification, solar research, liquefied coal and other synthetic fuels, geothermal energy, and other alternative energy sources.

Basically the business we are in is energy for tomorrow.

Gulf

Gulf people: energy for tomorrow.

Gulf Oil Corporation

Figure 7.13: Hydrogen From Water
This advertisement by Gulf Oil appeared in numerous popular publications in 1984, including *Time* and *Newsweek*.

In order to place the water issue in perspective, it is important to realize that it takes about 18 gallons of water to make a gallon of gasoline from crude oil. This is because oil refineries use large amounts of water to generate high-temperature steam, which is reacted with the crude oil to break or "crack" its long-chained hydrocarbon (hydrogen-carbon) molecules into the lighter molecules that make up isooctane fuels such as gasoline and aviation kerosene.

Water is a molecule that is made up of two atoms of hydrogen that are electrochemically bonded to one atom of oxygen. It is a relatively simple matter to separate the hydrogen and oxygen in water with electricity, which is a process referred to as electrolysis. It is also possible to separate the molecules of water thermochemically, that is, with relatively high temperatures (usually in excess of 1,600 degrees Fahrenheit). Such high temperatures can be achieved with both nuclear and solar sources, as the Gulf Oil advertisement in Figure 7.13 indicates. What is especially noteworthy about Figure 7.13 is that it demonstrates that a major oil company actually spent the resources to purchase full-page advertisements in a number of major news magazines to promote the concept of using hydrogen as fuel.

Most oil companies like to view themselves as not just oil companies, but energy companies. Oil companies have supported hydrogen technical conferences organized by a number of professional engineering societies, such as the International Association for Hydrogen Energy and the National Hydrogen Association. This is primarily because hydrogen can also be made from nuclear sources, and many oil companies have made substantial investments in nuclear fuels. But in any case, while many people automatically assume that oil companies will try to oppose the transition to a renewable resource "hydrogen economy," the reality is quite different. Companies like Shell Oil (which now has a Shell Hydrogen division), British Petroleum and Texaco have, to a certain extent, already been involved in researching and developing various solar and hydrogen production systems and when the governments of the world mandate that only renewable resource technologies will be acceptable, those will be the systems that the energy companies develop.

Although thermochemical methods of splitting water are promising, a far simpler method that has been used for years involves the process of electrolysis whereby an electrical current is passed through the water in a device called an electrolyzer. The electrolysis of water involves placing two electrodes, one positive and one negative, into a solution of water that has been made conductive by the addition of an electrolyte such as potassium hydroxide or sulfuric acid. As electricity is applied, the water molecules will separate, with the hydrogen gas molecules being attracted to the negatively charged cathode while the oxygen gas rises out of the solution at the positively charged anode. Water is continuously added to replace the water that has been broken down into hydrogen and oxygen.

To generate enough hydrogen to make the U.S. energy independent, about 4 million-acre feet (maf) of water would be required each year. An acre-foot contains 325,851 gallons, and although 4 maf is less than one percent of the annual flow of the Columbia River, which separates Oregon from Washington, it is almost what the entire state of Arizona consumes annually. While desert areas tend to have plenty of the land and sunshine that are required for solar technologies, the one thing they do not have is water. Thus, any proposal for a hydrogen production complex must have a realistic strategy for securing the necessary fresh water from sources other than existing rivers, aquifers, reservoirs or precipitation.

Water Options

While conventional fossil fuel and nuclear facilities require large amounts of water to produce electricity (about one to two gallons per kilowatt hour of electricity generated), solar or wind powered energy production systems will be able to produce electricity without any water requirements. As a result, the most obvious solution would be to make electricity in areas where the renewable resources are abundant, and then send the electricity over the utility grid system to where the feed-water for hydrogen production is more abundant. Assuming the electricity that is generated by wind systems in Nevada or New Mexico is sent to seawater desalination facilities located in southern California or Mexico, the large-scale production of hydrogen could then take place in centralized electrolysis facilities. Although this option is technically feasible, it is also possible to pump the desalinated seawater to where the electricity is being manufactured.

In the case of solar Stirling systems, they are able to incorporate an advanced -- but highly efficient -- high-temperature electrolysis process that has been under development by investigators at the Brookhaven National Laboratory and Siemens. This process is able to utilize part of the heat generated from the focal point of the solar concentrator to help break the water down into hydrogen and oxygen.

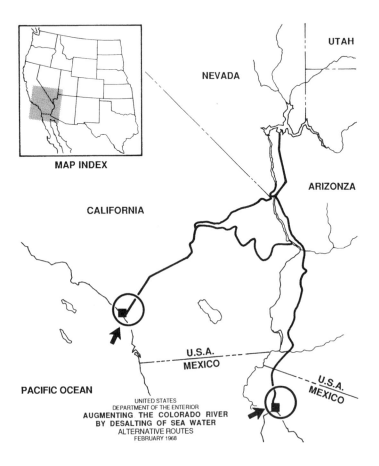

Figure 7.14: Freshwater Pipelines
A proposal developed by the U.S. Department of the Interior to pump desalinated seawater into the desert Southwest.

The use of high-temperature electrolysis can increase the efficiency of the conversion of electricity to hydrogen to nearly 100 percent. This is why there are substantial economic incentives to take advantage of the 1,200 to 1,500 degree Fahrenheit "fireball" of hot air that is generated at the focal point of a point-focus solar concentrator system. As such, pumping the distilled seawater to the electricity production centers may turn out to be the most cost-effective method of producing base-load quantities of hydrogen. While electricity would be needed to pump the water from the coastal areas to where it will ultimately be needed, it is presumed that the pumping stations would also be using electricity generated by a network of modularized renewable energy production units. However, there are other significant reasons why pumping large quantities of desalinated seawater inland is important. Chief among these reasons is the fact that fresh water is also at present a non-renewable resource that is being exponentially consumed by both urban and rural agricultural requirements for growth. The plan outlined in Figure 7.14 to bring large quantities of desalinated seawater into the arid desert regions was not developed by the U.S. Department of the Interior for hydrogen production, but for agricultural purposes. Fresh water aquifers are being seriously depleted in many parts of the U.S. and elsewhere, and like the energy crisis, there is no real plan of action to resolve the problem.

NAWAPA

One resolution to the Western water problem was proposed in 1964 by the Ralph M. Parsons engineering firm, located in Pasadena, California. Referred to as the North American Water and Power Alliance (NAWAPA), this project would involve diverting substantial amounts of fresh water (160 maf per year) from sources in Alaska and Canada where it is abundant (the annual runoff in these areas exceeds 800 maf), to areas in Canada, the lower 48 states of the U.S. and Mexico, where it is not. (Refer to Figures 7.15 and 7.16). Roughly half of the water would go to the U.S. under the NAWAPA proposal, and the remaining 80 maf would be equally divided between the countries of Canada and Mexico. The NAWAPA project would easily provide the fresh water necessary for the proposed modularized multi-quad solar-

Figure 7.15: NAWAPA
The North American Water and Power Alliance. Bringing water from
where it is abundant, to where it is not. The image was provided courtesy of the Parsons Engineering Company.

hydrogen production facilities in the Southwest. This would eliminate the need of having to pump desalinated ocean water up
from the Pacific Ocean. In addition, the NAWAPA proposal would
allow the Western river systems of the U.S. to maintain the critical salt balance necessary for the long-term survival of Western
agriculture. This serious salt accumulation problem has received
little public attention, but if it continues for another 20 or 30 years,
it will leave much of the richest agricultural region in the U.S. salt-
encrusted and barren.

Figure 7.16: NAWAPA Western Region
The image is from "The Salinity of Rivers," by Arthur F. Pillsbury who
had the article published in the July 1981 issue of *Scientific American*.

In a paper published in the July 1981 issue of *Scientific American*, Arthur F. Pillsbury outlined the salt problem in great historical detail. Pillsbury is especially well qualified to provide such insights. He served for many years as Chairman of the Department of Irrigation at the University of California at Los Angeles (UCLA). At the time of his retirement in 1972, he was professor of engineering and director of the UCLA Water Resource Center. In his paper, Pillsbury points out that many ancient civilizations rose by diverting rivers with a high salt content for the irrigation of their crops. Each one ultimately collapsed because as the water in the soil evaporated, it left behind the salt, which kept accumulating until the plants could no longer survive. The most fruitful of the ancient agricultural systems was the Fertile Crescent region. Six thousand years ago, the area was occupied by the Sumerians, who successfully diverted the river water to irrigate their crops. From there, civilization spread eastward through what are now Iran, Afghanistan, Pakistan, India and China.

At its peak of productivity each irrigated region is believed to have supported well over a million people. But all of those civilizations collapsed, and for the same reason: salt accumulation. Although floods, plagues and wars took their toll, in the end, civilizations based on irrigation faded away due to the accumulation of salt in their soil. Six thousand years later, the land in Iraq is still salt-encrusted and barren. If irrigated agriculture continues in the U.S., some of the richest farmland will also be lost.

All natural waters, including those described as fresh, contain salts and other dissolved minerals. According to Pillsbury, a virgin stream emerging from a mountain watershed may contain as little as 50 parts per million (ppm) of total dissolved solids of which salt is the major constituent. Ocean water, by comparison, has about 35,000 ppm. As the major rivers flow through arid regions for great distances, as is the case with all the major western rivers, the concentration of salts rises steadily as a result of evaporation. Rivers would normally carry the salts and other minerals dissolved out of rock and soil to the oceans, but when the rivers are diverted for irrigation, the salt gets trapped in the soil, where it keeps accumulating. The only effective means of flushing the salt out of the soil so it can continue its journey to the oceans is to irrigate the land with fresh water. The problem is that the fresh water that could accomplish the task is in the northern portions of Alaska and Canada.

This vast quantity of fresh water (about 800 maf/year) is not presently being used by anyone. It just keeps flowing into the Arctic Ocean where its fresh water status is terminated. NAWA-PA would be able to bring this fresh water to where it is critically needed -- but there are significant obstacles to implementing a NAWAPA-type water project.

One of the major environmental problems of the NAWAPA proposal is that it would involve constructing a large number of hydroelectric power plants that would be needed to provide the electricity to the pumping stations along the way. Hydroelectric facilities cause extensive damage to natural ecosystems that are affected. The hydroelectric facilities in the initial proposal would be able to generate an electrical surplus of about 70,000 megawatts, which is equal to about 20 percent of present U.S. electrical production. However, 70,000 megawatts is only equiv-alent to about 1.5 quads per year, which is only about 1.5 percent of the total amount of energy that is now consumed in the U.S. annually. This being the case, it would probably make more sense to eliminate virtually all of the hydroelectric facilities and use pipelines to deliver solar or wind-generated electricity or hydrogen that could then be used in the electric pumping stations. Even if the hydroelectric dams are eliminated from the original NAWAPA proposal, it still has significant economic obstacles to overcome. The cost of NAWAPA was initially expected to be $200 billion in 1965, which means the current cost could be in excess of $800 billion. Given these high costs, it is unlikely that a NAWAPA-type proposal will be implemented without substantial government assistance.

OTEC Sea Water Desalination

As discussed in Chapter 6, ocean thermal energy conver-sion (OTEC) facilities can be engineered to produce vast amounts of fresh water from the ocean. For example, a 100-megawatt OTEC power plant could minimally generate over 150 million gallons of fresh water daily -- as a by-product of the plant's electrical production process. If the OTEC plant were optimized for fresh water production, it could then generate up to 800 mil-lion gallons daily. The fresh water is produced by "flash evapo-rating" the warm seawater (i.e., it is vaporized at low pressure in a vacuum chamber). After the resulting vapor is expanded through turbines to generate electricity, it is then passed through cold-water condensers where the vapor (steam) will condense directly into pure water. Calculations have shown that the cost of fresh water from this process would be about $0.05 per thousand gallons at the plant site. Because the OTEC systems will probably be located about 600 miles offshore in the South Pacific Ocean, an additional cost of from $0.10 to $0.50 per thousand gallons will be required to transport the water by tanker or deep sea pipeline. However, even with these water transport costs, the fresh water from OTEC systems will still be less that half the cost of desalinating sea water by more conventional reverse osmosis, electrodialysis or distillation systems.

In comparing OTEC systems to NAWAPA, consider that if a 100-megawatt OTEC plant can generate 800 million gallons of fresh water daily, a fleet of roughly 93 such OTEC ships could provide the same amount of fresh water to the U.S. (i.e., 80 million acre feet per year) as the entire NAWAPA proposal. Assuming the OTEC plants were to cost about $100 million each, the total investment would be about $9.3 billion, which is only about 2 percent of the original $200 billion NAWAPA proposal estimate. Regardless of which water option that is ultimately selected, such long-term capital intensive macro-engineering projects will not happen without strategic planning and cooperation between governmental agencies and private industry. As of yet, no such cooperative planning effort exists or is anticipated. There is no question about the destructive nature of this lack of cooperation and planning. The only question is: *Is there still time to change course?*

Conclusions

This chapter has documented that there are more than ample renewable resources to provide for the energy needs of the human community. Point-focus solar Stirling systems (which are similar to cars from a manufacturing perspective), offshore wind power systems (which are similar to airplanes), and ocean thermal systems (which are similar to ships) are some of the most promising options that have thus far been developed. It is also worth noting that these three systems were developed more than 100 years ago. In addition, there are many other renewable energy technologies that could promise to be cost-effective methods of producing hydrogen in the future. These include photovoltaic cells, ocean current, wave, and biochemical energy systems. One the least expensive methods of producing hydrogen may come from the pyrolytic (i.e., decomposition by heat) or microbial (biological) conversion of biomass feedstocks, such as sewage sludge, paper and garbage.

In the final analysis, marketplace economics will determine which renewable energy technologies will be developed for large-scale electricity and/or hydrogen production. However, existing energy accounting systems do not take into account the external environmental and social costs that are associated with using fossil and nuclear fuels. Thus, before marketplace incentives will work, a "fair" accounting system will need to be established and used. As part of a legislative package, a suggested "Fair Accounting Act" that could be passed by both the federal and state governments will be outlined in Chapter 9.

Before the private sector industries will commit themselves to a long-term and costly energy transition, there will have to be a clear mandate from both federal and state officials. As with the related water problems, the primary obstacles are not technical, they are political. This underscores the fact that what is preventing such a political mandate is the general lack of awareness and communication between scientists who tend to operate in highly specialized fields; between elected officials who formulate energy and natural resource policies; and most of all, between the news media and the voting public who have the power to influence their elected officials. One can only hope that this communication problem is resolved before irreparable damage is inflicted on the Earth's biological life-support systems.

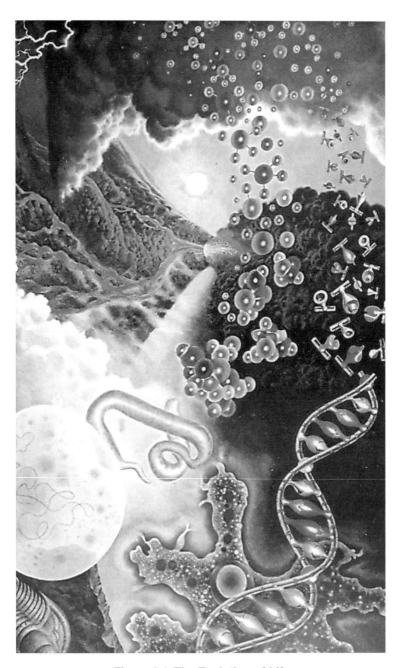

Figure 8.1: The Evolution of Life

Chapter 8

UTOPIA:
FROM HERE TO ETERNITY

"We are on a threshold of change in the universe comparable to the transition from non-life to life."

Dr. Hans Moravec
Director of Robotics Institute
Carnegie-Mellon University

This generation stands at the threshold of unprecedented developments. The great irony is that both utopia and oblivion are evolving simultaneously. If oblivion is defined as the contamination and destruction of the earth's biological life-support systems, then utopia could be defined as humanity having made a successful transition to renewable energy and biological resources. At present, there are very few people who realize that medical researchers are on the threshold of an era of molecular medicine that will make a biological transition to renewable resources inevitable. The ability to understand and change the genetic and amino acid codes that make up the molecules of life will surely be one of the most significant developments in human history, if not in the history of life itself.

Designer Genes

Molecular biology is a relatively recent field of science that deals with the chemical and physical composition and activities of the molecules that make up living organisms. Carl Sagan made a statement in his *Cosmos* television series that the chemistry of the relatively simple non-living molecules could be summed up in one or two paragraphs, whereas the chemistry of living molecules would take many volumes. This is because non-living "inorganic" molecules are usually made up of relatively few atoms that are bonded into a somewhat simple structure. In contrast, living "organic" molecules are usually made up of tens of thousands or millions of atoms that have been bonded in highly complex three-dimensional structures.

Research into the molecules of life is already having a profound global impact on medicine and medical research. Moreover, given the exponential explosion of the understanding of science in general, and biotechnology and computer science in particular; the stunning rate of progress is also accelerating exponentially. As a result, it has been estimated that more will be learned in the next ten years of medical science than has been learned from the beginning of time until now. That points one to the ultimate question: *Where is all of this leading?*

Perhaps the most important consideration is that medical researchers are on the threshold of a new era of "designer genes," that will allow individuals to select -- at will -- their genetic structure and code with atomic (i.e., atom by atom) precision. While investigators are currently focused on repairing or replacing defective genes that cause disease, in the not too distant future, it will be possible for people to take the next logical step in this genetic technology and choose their physical characteristics, such as hair color, physical size, or their sex. Such developments will have profound effects on legal, medical, and other social systems. These far-reaching impacts are inevitable because technology has a life of its own; and barring a collapse of the Earth's food production systems, the stunning advances in computers and biotechnology will continue. One of the spin-offs of such technology will be to greatly extend human life spans. This will primarily be accomplished by eliminating genetic and other molecular-based diseases, which includes the one fatal disease that effects everyone: the biological process of aging.

There are already five different peer-review scientific journals that are dedicated to understanding and defining the biological mechanisms of aging, and as the advances in molecular
biotechnology continue, increasing numbers of investigators will
begin to focus their attention on this ultimate degenerative disease that affects everyone.

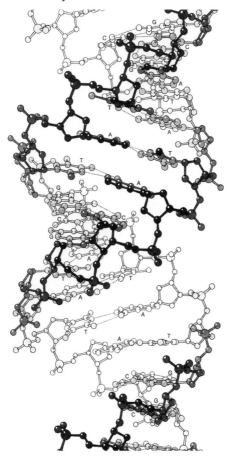

Figure 8.2: DNA Helix
In human beings, the DNA molecule is make up of over 100,000 genes
that contain a four digit code that provides instructions for the manufacture of proteins, which then determine all of a person's physical characteristics.   Twenty of the most common diseases that kill about 80 percent of the population are linked to defective genes, that can now be
removed or replaced by the rapidly evolving science of "gene therapy".

Blue Gene

The extraordinary advances in computer technology in recent years have served as the basis for many of the stunning advances in molecular biology and biotechnology. In 1998, IBM matched one of its supercomputers, named "Deep Blue" against Garry Kasparov, who was the world's chess champion, and Deep Blue won.

In the summer of 2000, IBM announced that it built Deep Blue's successor, which will become the most powerful supercomputer in the word. The new machine will be able to perform over 12 trillion operations per second, which is three times faster than today's existing supercomputers. This latest machine by IBM is intended to continue the advance toward matching and eventually surpassing the computing capabilities of the human brain. The new supercomputer is called "Advanced Strategic Computing Initiative White (ANCI White), it weighs 106 tons and it requires nearly 10,000 square feet of air conditioned floor space, which is equal to two NBA basketball courts. IBM will deliver ANCI White to the U.S. Department of Energy and Lawrence Livermore National Laboratory, which intends to use the machine to simulate the testing of nuclear weapons.

IBM officials indicated that the ANCI generation of super computers can also be used for a wide range of tasks. The new system will be expected to contribute to breakthroughs in financial models, genetic and molecular engineering, or reducing the time needed to create a global weather model from 18-hours to a few seconds. A single ANCI-class supercomputer could also monitor the air space of the entire United States.

IBM's ASCI Blue Pacific is currently the world's fastest supercomputer, which can perform approximately 3.8 trillion operations per second. According to Lawrence Livermore investigators, to perform a full three-dimensional nuclear simulation in 1985, it would have taken the fastest supercomputer available 60,000 years. In order for the machine to accomplish the calculations in a month, speeds of 100 trillion operations per second would be needed. IBM has also announced that it is now investing over $100 million over a five-year period to develop such an advanced supercomputer. This next generation machine will be called "Blue Gene," because it will primarily be used for biomedical and related protein engineering research.

Blue Gene will be able to operate 500 times faster than today's current generation of supercomputers and roughly two million times faster that a desktop PC. With speeds of over one quadrillion operations per second, Big Blue will be the first "petaflop"-class computer. According to a recent interview with Jared Sandberg in the December 13, 2000 issue of *Newsweek* magazine, Monty Denneau, one of the IBM researchers who is developing the Blue Gene architecture, if the speed of a conventional PC were measured on a bar chart as one inch tall, the bar for Blue Gene would be 30 miles! Paul Horn, the director of research for IBM, was quoted as saying "No one's dumb enough -- or smart enough -- to try to do this but us."

In order to provide Blue Gene with such scorching petaflop speeds, IBM needs to create an entirely new computer architecture. The machine will use over one million processors, compared to 5,000 processors that are used by existing supercomputers. But rather than forcing each processor to communicate with the memory system through a relatively slow communications link, IBM is imbedding the memory into the processor itself. In addition, the processors are relatively small and simple so that they can avoid the power-hungry data storage systems known as caches, and each of the processors will simultaneously handle eight separate series of computer instructions, known as threads. Finally, IBM researchers are convinced that the communications between chips will reach the kind of speed that could cover the Web's entire contents in about one second.

Why are such speeds needed? IBM hopes to solve one of the greatest mysteries of molecular biology: how the proteins our genes produce fold into their three-dimensional structure. The three-dimensional structure of nanobial proteins is the key to their chemical behavior. Given the critical nature of proteins to metabolism, understanding their folding process of is every bit as significant -- if not more so -- than understanding the human genome. However, the complexity of the task is so confounding that today's supercomputers can only simulate a small fraction of the protein folding process.

During the epic chess match, Kasparov uttered that Deep Blue momentarily played *"like a god."* Indeed, if Deep Blue's descendent, Blue Gene helps to solve the molecular mysteries of life, the work of God will be revealed with a level of detail that could not have even been imagined a generation ago.

The BioChip

One of the critical components in the biological transition to renewable resources will be the development of the "biochip," a molecular- based computer chip that will be fabricated not from silicon (like today's computer chips), but from the carbon and hydrogen that comprise the three-dimensional amino acid molecules that make up proteins. A typical example of the protein architecture (that of Cytochrome c) is pictured in Figure 8.4. In a paper that was published in *Scientific American* in March 1995, professor Robert Birge, director of the W.M. Keck Center for Molecular Electronics at Syracuse University, protein-based computer chips can potentially serve as computer switches because their atoms are mobile and change position in predictable ways.

The concept of using protein-based molecules to develop biochip capabilities was first proposed in 1974 by two IBM investigators, Arieh Aviram and Philip Seiden, and Mark Ratner, a chemist at Northwestern University. In 1983, the National Science Foundation (NSF) funded a technical conference on biochips that was also co-sponsored by The University of California at Los Angeles (UCLA). This "International Conference on Chemically Based Computing" was attended by leading international biologists, chemists, computer scientists, and electrical engineers, including the Nobel Prize-winning physicist Richard Feynman, of Caltech, who gave a talk on quantum mechanics and computing. The objective of the conference was to assess where the biotechnology and computer developments were heading, and to assess how such advancements would impact the global marketplace.

Biochip systems are currently being developed by a wide range of laboratories in addition to the ones already mentioned, including the U.S. Naval Research Laboratory in Washington, D.C., the Applied Physics Laboratory at Johns Hopkins University (Loral, Maryland), and UCLA's distinguished Crump Institute for Medical Engineering. It is only a question of time before true biochip capabilities are possible, and the implications are profound. Unlike the simple binary switches that are used in the current generation of silicon-based computers, protein-based enzymes have the ability to operate as "intelligent switches," whereby they can use their three-dimensional architecture for storing and transmitting potentially vast amounts of information.

Moreover, because enzymes can exist in many different conformational states instead of just two, they are capable of graded responses. According to F. Eugene Yates, Director of UCLA's Crump Institute, *"Enzymes are capable of not only saying yes and no -- but maybe."* Given the collective exponential advances in fields like molecular biology and computer science and engineering, it is reasonable to assume that if one is fortunate enough to live for another 20 or 30 years, one could be able to take advantage of the biochip technology to optimize their designer genes and the other molecular-based systems that will allow individuals to essentially eliminate the biological mechanisms of aging.

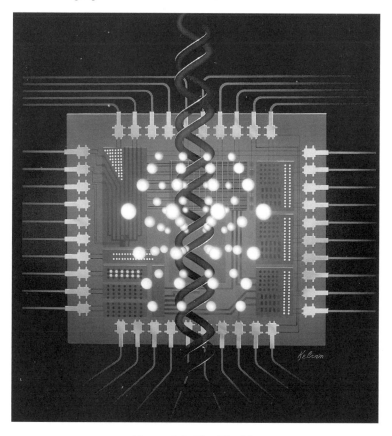

Figure 8.3: The Biochip
Integrating computers and molecular biology.

Immortalizing Enzyme

It is now known that cancer cells, which have been the subject of much investigation in recent years, are essentially immortal. Researchers now believe they have identified an "immortalizing enzyme" that is absent from normal tissue that allows cancer cells to divide indefinitely. Once investigators are able to modify the genetic instructions of this enzyme, it will then be possible to develop completely novel therapies that, unlike those available today, would leave patents largely unscathed. Huber Warner, a deputy associate director at the National Institute on Aging is one of the investigators in this field, and he commented on these developments in an article that appeared in the July 1994 issue of *Scientific American*. According to Warner, immortality is the norm among tumor cells and single-cell organisms. If the conditions are favorable, these cells can reproduce themselves forever. By contrast, normal human cells have a finite life expectancy. They may divide for a few generations, but they eventually stop reproducing and then die. An increasing number of cell biologists now suspect that the erosion of structures called "telomeres" is the reason.

Telomeres are specialized segments of deoxyribonucleic acid (DNA) that are found at the tips of chromosomes. They seem to stabilize the ends of the chromosomes and thereby prevent the chromosomes from degenerating. Molecular biologists often compare telomeres to the protective plastic caps on shoelaces. When a strand of DNA is duplicated during the process of mitosis, a few of the subunits at one end are always lost. As this process continues, the telomeres are whittled away. Eventually, the cells lose their ability to divide and they die.

Cancer cells and single-cell organisms are immortal because they are able to stabilize their telomeres. In the mid 1980s researchers demonstrated that protozoans make an enzyme, "telomerase," that adds new DNA sequences to the telomeres and thereby preserves their length. Experiments on human cells have shown that the enzyme telomerase is also used by both tumor cells and other single-cell organisms. Ironically, human cells also carry the gene for telomerase, but it is switched off after birth. The one exception in humans is the testis, which continue to use telomerase to rebuild the telomeres of sperm cells. If telomerase is an immortalizing enzyme it represents a significant opportunity for greatly improved cancer treatments.   Unlike conventional chemotherapy and radiation treatments, which damage all of the dividing cells in a patient's body, switching off or removing the telomerase enzyme will not damage the other healthy cells in the process. However, the ability to regulate telomerase will have potentially profound impacts on normal cells and the human aging process. According to Carol W. Greider of Cold Spring Harbor Laboratory in Long Island, New York, who has been conducting experiments on human cells, it should be possible to insert or switch-on the telomerase enzyme and thereby allow the normal healthy cells in the body to become immortal. If a person's cells are immortal, it follows that the person will also be immortal.

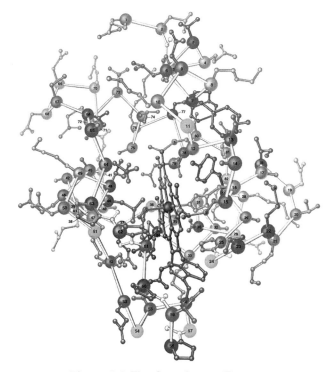

Figure 8.4: The Cytochrome Enzyme
The Cytochrome c molecule is wrapped in a protein chain that is made up of 82 amino acids. From "Cytochrome c and the Evolution of Energy Metabolism," by Richard E. Dickerson, Ph.D. Copyright March 1980 by *Scientific American, Inc.* All rights reserved.

Molecules of Memory

It has been said that a person without a knowledge of history is like a person without a memory. And a person without a memory is a person who could not possibly comprehend a reason for being. While our ability to memorize certain facts, and forget others is often taken for granted, it is the basis of our IQ and is an absolute prerequisite for literally everything that we do in our lives. It is hard to define the frustration by family members are not recognized by another member of the family that is afflicted with Alzheimer's disease, and it underscores the fact that when we define who we are, our memories serve as the basis of that definition of ourselves.

There are both long-term and short-term memories that are characteristically formed in humans. It is now known that if short-term memories are to become long-term memories, molecules within the brain's neurons where memories are made and stored must be physically and chemically changed. And as anyone knows, while long-term memory construction can occur in seconds as a result of a traumatic event in someone's life, studying for a chemistry or algebra exam typically involves a lot of hard work that is based on repetition. Moreover, some people seem to have a much easier time with long-term memory than others.

Jerry Yin and Tim Tully, who are investigators at the Cold Spring Harbor Laboratory have the distinction of first discovering a protein in the brain of fruit flies that is specifically involved in the formation of memories. In an article in the June 2000 issue of *Discover* magazine, Yin and Tully described how this protein comes from a family of proteins referred to as Cyclic AMP Response Element Binding proteins. In their experiments, Yin and Tully genetically modified one group of fruit flies to produce an abundance of the memory proteins and another group had little or none. They also utilized a control group of flies that had normal levels of the protein. The fruit flies were placed in glass tubes where they could be exposed to two different scented air currents. Both scents were attractive to the fruit flies, but one of the scents (licorice) was accompanied by an electric shock. A fly that is shocked every time it smells licorice will eventually build a set of neural connections that causes the fly to avoid licorice. What Yin and Tully wanted to know was how much repetition does is take for the long-term memory to form.

Yin and Tully tested the flies by placing them in the middle of a double-ended tube with the different scents that would be blown into each end of the tube. As the experiment demonstrated, flies with normal levels of the memory protein had to be shocked on an average of 10 times before they "learned" to stop moving toward the licorice -- just as a person must say a sentence over and over again to memorize the exact words. The flies that had little or none of the memory protein were never able to learn from their experiences, whereas the "smart" flies that were provided with an abundance of the memory protein only needed one electrical shock to learn their lesson.

On the basis of these experiments, Yin and Tully have been able to conclude that the extra memory protein significantly accelerated the process of memorization by eliminating the need for repetition. This finding could have profound implications for humans -- and any other animals that we might choose to enhance their memory construction capabilities. With the capabilities of genetic engineering and designer genes, it will soon be possible to have the body's brain cells increase the production of the of memory proteins. As a result, it will likely be possible for virtually anyone to have a photographic memory, and all that such a capability implies.

As the analysis of the three-dimensional structure of the molecules of memory continues, "smart pills' are inevitable. In the same way that people now load information from a CD onto the hard disk of their computer, in the not too distant future it should be possible to take an algebra pill or a chemistry pill that will contain all of the molecules of memory for each subject. With these evolving biotechnologies, intelligence and IQ will not be inherited as part of a random selection process of the genes, but selected with atomic precision by each individual. Thus, the designer gene era will not only allow people to select their physical appearance -- but their intellectual and emotional content as well. Anger, for example, is inherited in one's genes, and if there are people who choose to eliminate such genetic instructions, should they not be allowed to do so? While there are many people who will argue that it is only acceptable to use these biotechnologies to help physically disabled or mentally retarded people become "normal," the reality is that most people will want to have the option to live as long as they choose -- and not just as a human being -- but as a *biocybernetic organism.*

Biocybernetic Evolution

"Cybernation" refers to the integration of computers and machines, and the term "cybernetics" refers to the science that studies the similarities that exist between the electrical circuits of computers and the human nervous system. The development of the biochip "on-board" personal computer-companion will help to make the biocybernetic capabilities possible. In an era where aging, disease and death will have essentially been eliminated, most human beings, as we now understand the term, will have chosen to evolve into a new species -- if not a new life form -- as an "immortalist" culture of biocybernetic organisms.

The evolution of biocybernetic organisms will be the inevitable result of the synthesis of computer and biotechnologies that is expected to occur in about 30 to 50 years. This calculation is based on the fact that, on average, the cost of computation has been cut in half every 24 months from the time of the first mechanical adding machines that were developed nearly 100 years ago, to the current generation of supercomputers. As a result, there has been a trillionfold decrease in computing costs; and the rate of progress is increasing at a very rapid exponential rate. According to Hans Moravec, the director of the Robotics Institute at Carnegie-Mellon University, if the present exponential rate of progress continues, by the year 2010, computers will be a thousand times faster than they are at present, which will make them comparable to the human brain. By the year 2030, he believes the integration of biotechnology, computers and robotics will have produced a biocybernetic organism that will integrate biochip and designer genes into existing human beings.

Robert Jastrow, who is a physicist and professor of astronomy at Columbia University, believes human equivalence in computers will take place even sooner, perhaps by the year 2010. But even assuming Moravec's more conservative assumptions are correct, if human equivalence is possible by 2030, *what will be exponentially possible by the year 2040 or 2050*? At the very least, it is reasonable to expect that the process of natural selection will have been replaced by "biocybernetic selection." In his book, *Mind Children,* Moravec, argues that computer technology is advancing so swiftly that there is little we can do to avoid a future world run by super-intelligent robots. The real question is: *Will we become them?* Moravec is quoted as saying:

"In an astonishingly short amount of time, scientists will be able to transfer the contents of a person's mind into a powerful computer, and in the process, make him -- or at least his living essence -- virtually immortal."

Moravec is not alone in his beliefs. Gerald Sussman, a computer researcher at the Massachusetts Institute of Technology (MIT) and an author of a textbook on artificial intelligence, also believes that computerized immortality for people *"isn't very long from now."* Sussman argues, "If you can make a machine that contains the contents of your mind, then the machine is you." And while an individual machine may not be able to last forever, back-up copies could provide a large measure of insurance that the essence of the memory and/or personality would never be lost. Jastrow holds the somewhat logical view that the size of the human brain, and thereby the size of its memory storage area, has been inherently limited by the size of the female pelvis. He argues that offspring that had been genetically selected to have larger brains were unable to pass through the birth canal of the female pelvis, which often resulted in the death of both the mother and the child. Computer memory systems have no such size limitations. As a result, the projected long-term growth of computer memory systems will result in implications that are impossible to foresee at present.

One of the most fascinating and technically impressive visions of what is coming is detailed by K. Eric Drexler in his book, *Engines of Creation.* Drexler, who is a graduate of MIT, served as a National Science Fellow and a research affiliate at the MIT Space Systems Laboratory. He is currently President of the Foresight Institute, located in Palo Alto, California, and an instructor at Stanford University, where he currently teaches a course on nanotechnology. The term "nanotechnology" is defined as the manipulation of individual atoms or molecules to build more complex structures to atomic specifications. A nanometer is one-billionth of a meter, and such a scale is required if one is to engineer on an atomic scale.

As Drexler points out, nanotechnology and molecular technology are interchangeable terms that will lead to nanocircuits, nanomachines and nanocomputers, which will ultimately change our world in more ways than we can imagine. The basis of Drexler's argument is firmly rooted in the principals of molecular chemistry that microorganisms -- or more specifically, the enzymes and other protein molecules that manufacture and operate the microbes -- have been successfully using for over three billion years.

The molecular mechanisms of proteins have been under study for years by a large number of biochemistry investigators at distinguished academic centers in virtually every industrialized country. In the U.S., major biotechnology research programs have been underway for years at virtually all major universities, including Stanford, Harvard and MIT. It is worth noting that prior to 1970, the general consensus in the scientific community was that tampering with the molecules of life was clearly impossible. Then in 1971, scientists at the U.S. National Institutes of Health found that they could successfully transfer the genetic molecules of DNA from one organism to another by using a virus to do the actual work. Many other biotechnology investigators soon duplicated and improved upon the process that is now referred to as "recombinant" DNA technology. It did not take long before this powerful new research was utilized by private industrial research laboratories around the world, many of which are now working 100-hour weeks on biotechnology developments and potential products.

A confirmation of the stunning progress being made in biotechnology occurred in 1980 when the U.S. Supreme Court issued a landmark ruling that man-made organisms could be patented. The fact that, prior to 1970, the general consensus in the scientific community was that genetic engineering was impossible underscores the significance of the profound advances that have been accomplished in recent years. To understand the significance of these developments in molecular biology, it is helpful to have a basic understanding of the ancient microbial and protein-based organisms that are at the very heart of metabolism -- and life itself. Since microbes such as bacteria operate on the micrometer scale (and are therefore called microbes), and enzymes and other proteins operate on a nanometer scale, the protein-scale molecules would be nanoorganisms (i.e., nanobes). Given their seemingly incredible array of activities, the Earth's nanobial "founding fathers" appear to be a highly advanced civilization that is over 3.5 billion years old. The story of the nanobes is the story of life itself. While there is still much to be learned about these life forms, modern science has revealed a great deal about nanobe evolution.

The Creators

Prior to the invention of the microscope, microbes were unknown "spirits" that could bring good health -- or disease and death. When microbes were first observed in the 17th Century, they were considered mindless little "bugs" that just happened to be everywhere there was life. Creationists who believe that the human brain and body are simply too complex and sophisticated to occur by accident are at least partly correct. However, they may be surprised to find that this implies that the true image of our Creator may be more like that of the nanobial proteins pictured in Figures 8.4 and 8.5. There is no question that the complexity of the human brain, a single cell, or even the structure of an enzyme is extraordinary. But the human nervous system, which is the most complex structure in the universe, is the end product of billions of years of nanobial evolution, and it is directly operated and maintained by a current generation of nanobes who are the citizens within the city of the cell. Whether it is the subconscious thought one thinks, the food one eats, or the selective attraction one has toward another, there are trillions of nanobes that are the ultimate directors and recipients of these activities. In a biological context, we have been created by -- and live for -- the nanobial intelligence that is at the heart of life itself.

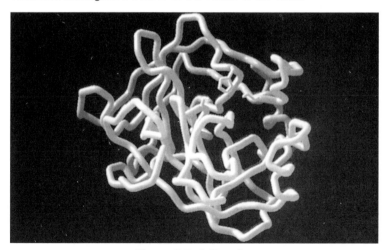

Figure 8.5: A Nanobe
The computer graphics was provided courtesy of Arthur J. Olson, Research Institute of Scripps Clinic, Copyright © 1985.

Make no mistake about it: these tiny organisms are responsible for manufacturing, operating, and maintaining virtually all
cellular organisms (including all green plants, animals and
humans) that exist on the Earth. Nanobes make microbes,
humans and other animals in the same way that humans make
large structures such as an aircraft carrier. And just as an aircraft
carrier is controlled by a few officers on the bridge, there are specialized microbial enzymes that occupy and regulate the human
command and control centers within the central nervous system.
Vast numbers of other enzymes are responsible for programming
and maintaining the DNA biocomputer that is responsible for storing the memory of amino acid structures and sequence codes for
all the different types of nanobial-scale proteins that make up living organisms.

In 1992, *Time* magazine ran a cover story article entitled,
"What Does Science Tell Us About God." Time journalist Robert
Wright provided an interesting overview of the subject, but he
completely missed the issues related to the nanobes and their
critical role in the evolution of life on the primitive Earth. Science
and religion do serve similar societal functions with respect to
providing an explanation as to how things got to be as they are.
In Genesis, biblical scripture indicates a "creationist" view whereby human beings and other plants and animals were created in a
matter of days by a human-like God.

Scientific inquiry provides a very different explanation that
begins roughly 15 billion years ago starting with the Big Bang.
While no one in the scientific community has yet proven exactly
how living organisms evolved out of non-living matter, it is clear
that the original life forms were not mammals or fully-formed
human being-type organisms, but rather molecular-sized proteins
that in a biological sense created, and continue to operate within
the genetic and cellular structure of human beings -- as well as
virtually all other living plants and animals. Even bacteria and
viruses are engineered by and operated for the nanobial-scale
proteins that are intimately integrated into their structure. This is
not a theory. Indeed, no credible biochemist would dispute these
basic points. Proteins (which include all enzymes) are at the
heart of metabolism. It is certainly reasonable to assume that the
human brain is far too complex a structure to have been simply
been created by random chance. A watch implies a watchmaker,
and in the case of living organisms on the Earth, the "Creator"

appears to be a vast and ancient organism that is more that 3.5 billion years old. This collective entity has engineered and mass-produced the most complex machine yet discovered: the human brain and nervous system. Given their seemingly incredible array of activities, the Earth's "nanobial founding fathers" appear to be a highly-advanced civilization that started with a single, original protein (Adam?), which figured out how to utilize ribonucleic acids (RNA) to reproduce itself (thereby producing a kind of Eve). Along the way, there were lots of opportunities for experimental changes that have eventually resulted in you and me, the latest in the multi-million year-old mammalian assembly lines.

Spiritual prophets in the past had no concept of the micro-bial or nanobial world. Electron microscopes and scanning tun-neling "nanoscopes" would not be invented until the 20th Century. Although nanobial enzyme reactions were first observed in the 1830s, little was known about their structure and evolution prior to the 1970s. This was because the surface imprints of bacterial fossils provided little information about their origin or what their internal molecular structure could have been like. However, with recent advances in molecular biology, scien-tists are no longer limited to just analyzing the geological fossil record. It is now known that by carefully analyzing the amino acid sequences within a cell's proteins and the nucleotide sequences of its nucleic acids (i.e., DNA and RNA), it is possible to provide a remarkably accurate picture of the cell's genetic origin. This molecular record yields three kinds of evolutionary information: time; genealogical relations; and a record of ancestral character-istics. This living record provides far more information than geo-logic fossils because it provides a detailed picture of the evolution of the internal molecular structures of the organisms. Moreover, the living molecules reach back to a time long before the oldest fossils, to a period when the common ancestor of all life existed.

When people examined the human body in ancient times, they could identify major organs that they could see with their eyes, but they could never find the "Spirit" or "Soul" that was con-sidered to be the ultimate life force within people. It was, in fact, the molecular-scale organisms that could bring good health -- or disease and death. Given the invention of microscopes and the current scientific understanding of the molecular world, it is now clear that the nanobial proteins are indeed the essential require-ment for life, for no living organism can survive without them.

Nanobial Origins

Microscopes have revealed a microbial universe within living cells that ultimately provides a pathway to the nanobial world. This is because it has now been well established that virtually all of the biochemical reactions necessary for life could not take place without the nanobial enzymes who manage and maintain the molecular machines of life. Even the DNA code, in all its complexity, contains the precise instructions to manufacture the wide-range of nanobial proteins and/or enzymes that are an essential prerequisite for metabolism.

While evolution (which simply means change) is generally regarded as an observable fact (i.e., each offspring has a unique genetic combination of its parents and is therefore different from both), how life originated (i.e., where did the first nanobes come from?) is still very much a matter of speculation. One intriguing explanation, referred to as the "clay theory," is a more refined explanation of the traditional "primordial soup" theory.

The traditional theory holds that amino acids were synthesized as a result of having the primordial elements of hydrogen, carbon, oxygen and nitrogen "cooked" by solar energy for roughly 500 million years. The clay theory provides a potentially significant insight into how such organic molecules may have initially evolved on the primitive Earth. Lelia Coyne, one of the senior scientists at NASA's Ames Research Center in California, believes that clay particles may have indeed served as the necessary catalyst for the evolution of living molecules. This is because clays have the capability to store energy (in the form of energized electrons) for later use in chemical reactions.

The crust of the Earth is rock, but as it breaks down into smaller fragments and particles it evolves into soil, sand, silt and clay. Clays are the smallest and finest of the mineral particles, and they are usually made up of silicate crystals that can take the shape of well-formed hexagonal structures. Such crystals could have served as the template upon which the biological molecules could form their similar hexagonal structures (see Figures 8.6 and 8.7). Graham Cairns-Smith of Scotland's University of Glasgow suggests that clays are "proto-organisms" that could have served as the pattern for living systems. He notes that metals in clay lattices can form complexes with the precursors of proteins and DNA.

These lattices provide the molecular structure that is necessary to store the energy needed to catalyze chemical reactions, and most importantly, to reproduce. In a paper titled "Clay" published in the April 1979 issue of *Scientific American,* Georges Millot, who is a member of the French Academy of Sciences and professor of geology at the Universite Louis Pasteur in Strasbourg, also confirms the unique electrochemical properties and geometric structure of clay. Millot believes these specialized properties may have played an important part in allowing amino acids to evolve into the long peptide chains that make up one of the key biological structures: the proteins. It is now well established that protein molecules are indeed the building blocks of life. Depending upon their configuration, proteins can become the structural material from which living tissue is made; they can become the hormones that regulate all chemical behavior in the body; or they can become the all-important enzymes that are critical to life -- because they mediate virtually all of the biochemical reactions in living organisms. Proteins store our memories, and they represent a critical link in the origin of life. Indeed, the protein molecules stand at the very threshold of life where chemistry becomes biology. They are, in a biological sense, our creators.

This remarkable range of functions results from the folding of proteins into distinctive three-dimensional structures that can bind highly diverse molecules. A major goal of biochemistry is to understand this highly complex protein folding process, and how amino acid sequences that make up DNA specify the protein's conformational structure. It was the nanobial proteins that originally created and continue to maintain the DNA biocomputer, which stores the amino acid codes for their structure. The image in Figure 8.5 is the nanobial enzyme subtilisin, but there are more than five thousand different types of enzymes in the body, and each enzyme has a unique molecular structure and appearance.

Nanobes Evolve Into Microbes

With time, the nanobial civilization refined its molecular machines into increasingly sophisticated structures. A key milestone in their progress was the development of vastly larger and more complex organism that could contain literally thousands of highly specialized nanobes -- *the Microbe*.

Figure 8.6: Clay: A Catalyst for Life
The image is reprinted from "Clay" by Georges Millot.
Copyright © April 1979 by *Scientific American*. All rights reserved.

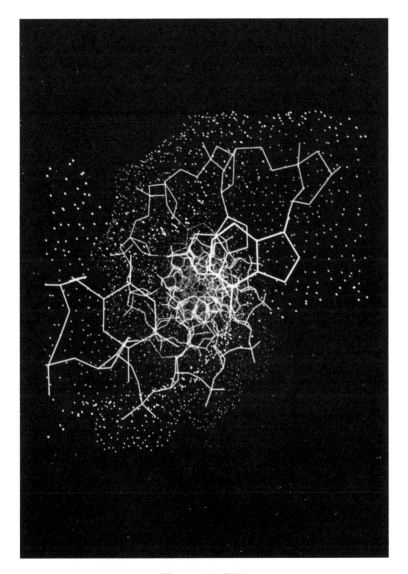

Figure 8.7: DNA
A computer simulation of what it would be like to look down a segment
of the molecular structure of DNA. The genetic sequence yields precise
information about the evolution of molecular structures. The living molecules reach back to a time long before the oldest fossils, to a period
when the "common ancestor" of all life existed.   From the cover of
Scientific American, October 1985, by Arthur Olson, Ph.D., Department
of Molecular Biology, Scripps Clinic & Research Institute, La Jolla CA.

An excellent overview of the microbial world was provided in the November 8, 1999 issue of *U.S. News & World Report*, titled "The Invisible Emperors," by Charles Petit and Laura Tangley. In their article, Petit and Tangley make the following observations:

> *"Earth's microscopic hordes make the rest of life possible; their sheer mass, vitality, and variety dwarf that of the readily visible world. Most are not only harmless but oblivious to people, who cannot survive without them. Increasingly, however, humans are paying attention. Scientists are mining the microbial world for potential cancer cures; one lowly microbe helped spawn a revolutionary technique to replicate DNA. Research on microbes is even changing our understanding of the nature of life -- and perhaps its origins. The famed Harvard University evolutionary biologist Edward O. Wilson, who spent much of his career exploring the intricate worlds of ants, says that if he were starting over today he would study microbes: "The microbial world is without limit." Without microbes there would be no food to eat or air to breathe: Algae and bacteria, through a process called photosynthesis, produce up to half the atmosphere's oxygen. . . Natural residents of the human body, microbes help digest food, produce essential vitamins, and protect us from other organisms that cause disease."*

James Shapiro, professor of microbiology at the University of Chicago, summarized his view of the microbial phenomenon in a June 1988 paper published in *Scientific American:*

> *"Without bacteria, life on earth could not exist in its present form. Bacteria are key players in many geochemical processes, including the fundamental nitrogen, carbon, and sulfur cycles, which are critical to the circulation of life's basic elements. If these processes were to grind to a halt, the planet's soils, waters and atmosphere would become inhospitable for life. Yet in spite of such global importance, the notion has persisted that bacteria are simple unicellular microbes."*

That view is now being challenged. Investigators are finding that, in many ways, an individual bacterium is more analogous to a component cell of a multicellular organism than it is to a free-living, autonomous organism. Bacteria have been found to form complex communities, hunt prey in groups, and secrete chemical trails for the directed movement of thousands of individuals. Although bacteria are tiny, they display biochemical, structural and behavioral complexities that outstrip scientific description. In keeping with the current microelectronics revolution, it may make more sense to equate their small size with sophistication rather than with simplicity.

Utilizing molecular methods of analysis, a view of microbial evolution has been provided by Carl R. Woese in his paper "Archaebacteria" published in the June 1981 issue of *Scientific American*. Woese received a Ph.D. in biophysics from Yale University in 1953 and is currently professor of microbiology and genetics at the University of Illinois. He has devoted much of his professional life to understanding the evolution of the thousands of cellular ribosomes that are responsible for translating and pro-cessing genetic information, which is then used to manufacture specific proteins. As Figure 8.8 shows, archaebacteria, eubacte-ria and eukaryotic (i.e., nucleated) cells represent the three pri-mary kingdoms of microbes that are thought to have evolved from a common microbial -- and ultimately nanobial ancestor.

Woese points out in his paper that early natural philosophers believed that all living things were either plants or animals, and when microbes were discovered, they were divided into the same classification. As microscopic investigations continued, it soon became apparent that the plant or animal classifications would have to be expanded to include fungi, protozoa and bacteria. With the development of more sophisticated analytical tech-niques that involved the direct sequencing of DNA and RNA mol-ecules, Woese and his colleagues were able to establish a detailed picture of microbial evolution that is summarized in Figure 8.8. Woese points out the following:

"Perhaps the most exciting thing about the recent dis-coveries in molecular physiology is that they show how much information about the very early stages of evolution is locked into the cell itself. It is no longer necessary to rely solely on speculation to account for the origins of life...

...It has become customary to think of the last decades of this century as a time in biology when 'genetic engineering' will make possible exciting developments in medicine and industry. It must also be recognized that biology is now on the threshold of a quieter revolution, one in which man will come to understand the roots of all life and thereby gain a deeper understanding of the evolutionary process."

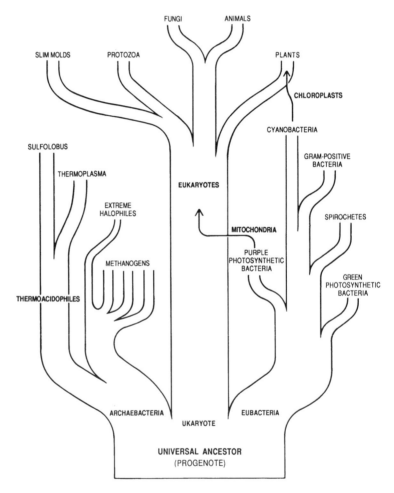

Figure 8.8: Microbial Evolution
From "Archaebacteria," by Carl R. Woese.

Microbial Memory

While human beings have a well-developed consciousness, and animals such as dogs and cats obviously experience some level of consciousness, smaller animals, such as insects, are viewed as essentially non-intelligent mini-robots that would not be expected to have a consciousness any more than the family automobile. Observing a common house roach running for its life should convince one that such insects are far more conscious than inanimate objects such as cars. Most people, however, are usually trying to swat the roach rather than carefully observing its sensory response to danger. Microbial and nanobial organisms are much smaller than insects, and as a result, it is generally assumed that such tiny organisms could not possibly have thought processes. Indeed, looking at such creatures under a microscope gives no hint that they could be aware of themselves or their environment. To put this issue into some sort of non-anthropomorphic perspective, time-lapse filmmakers have set up cameras at busy intersections or shopping malls that would take pictures of events as they were happening. As part of a special effect, when the film was shown, it was speeded-up so that the automobiles began to look just like masses of red blood cells surging through arteries, and the individual people scurried about the shopping malls just like the microbes under the microscope. By speeding-up the film, the whole character of the subjects under study changed dramatically.

Few people have attempted to look at the world through the perspective of a microbe -- much less that of a nanobe -- but for those who have, remarkable insights have been obtained. One of the first investigators to do so was a German biologist, Wilhelm Pfeffer, who reported his observations in 1883. He had filled small tubes with mixtures of repellents and attractants and noticed that bacteria would swim into a tube after the attractant, even though the repellent by itself would have sent the bacteria scurrying in the opposite direction. What the bacteria did is similar to someone braving a buzzing beehive to get some honey. Such actions imply analysis of stimuli followed by a decision to take action. Pfeffer's observations essentially went unnoticed until 1974, when two microbial biochemists, Julius Adler and Wung-Wai Tso, prepared a paper for *Science* magazine titled: "Decision-Making in Bacteria."

Paul Pietsch, a neuroanatomy professor at Indiana University, described the research of Adler and his colleagues who used strains of escherichia coli bacteria whose genes had been thoroughly studied to identify specific molecules that could attract or repel individual microbes. It did not take long for Adler to identify and isolate the individual proteins responsible for acting as the microbes' sensory receptors. Daniel Koshland, a biochemist at the University of California at Berkeley, refers to such molecules as the bacteria's "eyes and ears." In addition to these sensory systems, other protein molecules have been identified as data processors -- a kind of bacterial brain -- that is involved with the storage of microbial memories.

Koshland has indicated that the bacterial nervous system is actually sophisticated enough to discriminate between one part in 10,000. This is equivalent to a human detecting the difference between a jar with 9,999 pennies and one with 10,000. Since few, if any, humans could make such a distinction, this remarkable observation implies that in some ways, the microbial mind appears to be capable of performing functions the much larger human nervous system cannot. In explaining how such powerful computational skills could be packaged in an organism as small as a microbe, one needs to understand how three-dimensional images called "holograms" work.

Pietsch, who authored the book *Shufflebrain: The Quest for the Hologramic Mind,* explains that holograms are able to store images by recording the relative positions or phase relations of light waves. Phase is built into all cyclic events, from light waves to waves in the ocean. Thus, just as a ripple in a pond can have the same profile, phase relationships can be recorded in an area that is very large or microscopically small. This is because it is the pattern of interaction between waves that is detected. If memory functions on a similar principal, our microbial -- or nanobial -- companions may be more intelligent than we could have suspected. Pietsch explains the specific mechanisms as follows:

"On the sensory side, 30 kinds of chemical receptors and a dozen or so proteins that function as data processors lie at or near the inner surface of the cell's membrane. A receptor, activated by a stimulus, fastens itself onto a data processor. The processor, in turn, signals two oppositely acting enzymes in the cell membrane.

These enzymes chemically change the proteins near the flagella in the cell membrane by either adding or removing a small chemical unit called a methyl group. Each time the group is added, protons (i.e., hydrogen nuclei) are released to run the flagellar motor. For those in search of the microbe's memory, there is a very interesting fact about the addition and removal of the methyls. The enzyme systems that perform the two tasks operate at different rates. Those that put on the extra group are faster than those who strip it off. Inside the cells, there remains a microscopic chemical record -- a memory -- of events that occurred a few seconds in the past. In the cycle changes of methyl there is a phase lag, and phase lag allows for hologramic memory."

The concept of using holographic storage systems is also under investigation by a number of computer manufacturers. This approach was discussed in some detail in an article, "Avoiding a Data Crunch," by Jon Toigo that was published in the May 2000 issue of *Scientific American.* Toigo points out that the technology of existing computer hard drives is fast approaching a physical barrier imposed by what he refers to as the superparamagnetic (SPE) effect. Simply described, SPE is a physical phenomenon that occurs in data storage when the energy that holds the magnetic spin in the atoms making up a bit (either 0 or 1) becomes comparable to the ambient thermal energy. When this happens, the bits become subject to random switching between 0s and 1s, thereby corrupting the information they represent. With the current pace of miniaturization, some experts are predicting that the industry may hit the SPE wall as soon as 2005. Holographic memory systems, which have been in the research and development stage for more than three decades, have the potential to eliminate the SPE barrier. According to Toigo, holographic technologies permit retrieval of stored data at speeds not possible with the existing magnetic methods of storage. Indeed, no other storage technology under development can match holography's capacity and speed potential.

Because of the potential advantages of holographic memory systems, a number of companies, including IBM, Rockwell and Lucent Technologies, are developing such systems in cooperation with the U.S. Defense Advanced Research Projects Agency

(DARPA). Since the mid-1990s, DARPA has funded two different groups that have been working on holographic memory storage systems: the Holographic Data Storage System (HDSS) consortium and the Photo Refractive Information Storage Materials (PRISM) consortium. Both of these teams bring together private companies and academic investigators at a number of major universities, including the California Institute of Technology, Stanford University, the University of Arizona and Carnegie-Mellon University. HDSS, which was formed in 1995, was given a five-year mission to develop and demonstrate a practical holographic memory storage system; whereas PRISM, formed in 1994, was commissioned to produce advanced storage media for use in holographic memories by 2001. While the quest for an ideal storage medium continues, it is clear that holographic memory storage systems are one of the most promising options currently under development.

If Pietsch and his colleagues are correct, the nanobes and microbes discovered this holographic memory storage principal more than three billion years ago. Memory implies intelligence, and intelligence implies an ability to learn. According to Koshland, if young bacteria do not encounter certain molecules early in their life, they will never develop the mechanisms to perceive those molecules later on. Thus, bacteria and other microbes seem to be capable of acquiring, as well as storing, information in their microbial nervous systems. But it is not the individual microbial or nanobial mind that is as significant as the potential collective effect of all of the Earth's trillions of nanobes that interact in virtually every metabolic activity of living organisms. Thus, all of the nanobes and/or microbes on the Earth may indeed collectively make up a global "superorganism."

This view of a microbial superorganism was first advanced in 1789 by a Scottish scientist, James Hutton, who formally introduced the idea at a meeting of the Royal Society of Edinburgh. Since that time, many scientists have elaborated on Hutton's superorganism theory, and recent understandings in molecular biology suggest that the countless trillions of microbes that now make up the superorganism on the Earth all originated (i.e., were descended or cloned) from a common nanobial ancestor some 3.5 billion years ago. If this is the case, it means there was an original nanrobe who figured out how to manufacture copies of itself, and thereby achieve a kind of biochemical immortality.

Given this view, it is improper to think in terms of a first or original human being, because what are defined as human-type organisms have been evolving for at least three million years. The primate ancestors of modern humans that did exist 3 or 4 million years ago would not be able to get into a restaurant without being noticed. However, it does appear as though there may have been an original nanobe, and that nanobe may still be technically alive. It is also possible that the original nanobe, or nanobes, may have evolved into a global superorganism that is made up of the sum total of the Earth's biota.

One view of the microbial superorganism was provided in a paper by Sorin Sonea, "The Global Organism," published in 1988 by the *New York Academy of Sciences*. Sonea is a professor of microbiology and immunology at the University of Montreal, in Quebec, and she is co-author of the book, *A New Bacteriology*. Sonea explains that bacteria have not only demonstrated an ability to communicate molecular changes around the world -- but they are apparently are able to do so with surprising speed. Sonea uses the example of penicillin, which was developed in World War II to help fight staphylococcus infection. But in less than 10 years, the staphylococcus had developed a resistance to penicillin, as well as five other antibiotics. Moreover, the resistance was worldwide. According to Sonea:

> ". . . this baffled biologists who believed that bacteria acquired resistance through random mutation. But such a time consuming process could not account for the speed with which the microorganisms had developed defenses against antibiotics. Then, in 1963, Tsutomu Watanabe, a Japanese bacteriologist, published a paper in Bacteriological Reviews that stunned the medical community. Watanabe showed that the sudden resistance of shigellae was due to genes supplied by virus like elements, called R plasmids, which are capable of replicating independently of the cells they occupy. On every continent, plasmids appeared and rapidly spread, migrating from one kind of intestinal bacterium to another. If ever there was an idea in need of revision, it was that bacteria are backward, easily managed creatures inhabiting the margins of existence."

Sonea explains that to study bacteria in isolation is like trying to understand the complexity of a human being by studying a single cell:

"The point of departure for this new approach is the realization that in nature there are no isolated bacteria; they all depend genetically and metabolically on one another. Indeed, all bacteria on the Earth contribute to, and draw benefits from, a common gene pool; together, they constitute the communications network of a single superorganism whose continually shifting components are dispersed across the surface of the planet. Like the billions of cells that constitute a human body, all the bacteria in the world derive from a single cell. To be sure, individual bacteria are dying all the time, as the cells in a body, but the ensemble of surviving bacteria -- the planetary superorganism -- continues to thrive without interruption."

This microbial superorganism view has also been advanced by James E. Lovelock, a British atmospheric chemist and technical consultant to NASA. Lovelock has specialized in developing the instrumentation to measure minute amounts of gases that exist in the atmosphere of the Earth, or any planets that are observable with a telescope. He is particularly interested in evaluating atmospheric gases such as carbon dioxide, methane, oxygen and other gases that are absorbed or generated in significant quantities by the sum total of the Earth's biological organisms. As a result of Lovelock's research, he concluded that the Earth's microbial superorganism, that he calls "Gaia," a Greek word meaning Goddess of the Earth, was not just riding along on Spaceship Earth like a passenger. Rather, this collective planetary Gaian life force was actively and purposely regulating the Earth's climate by the production and/or consumption of various atmospheric gases. Lovelock has summarized his hypothesis in his book, *Gaia: A New Look at Life on Earth,* and in a documentary interview produced by NOVA and the Public Broadcasting System (PBS). He points out the remarkable fact that although the Sun has increased its heat output by roughly 30 percent since the microbes first began populating the Earth some 3 billion years ago, the atmospheric temperature has been maintained at a relatively constant level of between 5 and 25 degrees Celsius.

According to Lovelock:

> *"It suddenly dawned on me that even more remarkable than the strange and potentially explosive atmosphere that we had, is that it remained constant, like that, over geologic periods of time. And this is an event as unlikely as riding on a bicycle, blindfolded, unscathed, through rush hour traffic. It is something totally impossible -- and it required the presence of some sort of automatic regulating system that could keep this constant."*

Lovelock is not suggesting that the Gaian superorganism is a conscious, intelligent, all knowing, God-like being that is actively involved in foresight and planning. Rather, he takes the view that Gaia is the sum of the microbial biota, and that it has an automatic temperature regulating system that can actively adapt (somewhat like a slow computer) to environmental changes. However, Lovelock is convinced that the existing microbial stable-state system on the Earth is rapidly reaching its limits to absorb carbon dioxide in the atmosphere, and the problem is being compounded by the destruction of the Earth's forests and the widespread use of fossil fuels. This means the human population and other mammalian life forms may soon be extinguished in the rapidly changing climactic environment, much in the same way that the dinosaurs suddenly disappeared after their 150-million year dominance. The major groups of nanobes will surely survive for a time by engineering new organisms that will be able to adapt to the more severe environment. Indeed, if the nanobial superorganism does turn out to be a purposeful, consciousness being, perhaps the biocybernetic organism is what is now on the nanobial drawing boards.

If one assumes that an individual nanobe organism, or the larger superorganism the individual nanobes collectively make up, is a highly intelligent entity; it should be possible to communicate with this ancient civilization that lives within us. To understand how this might be possible, one must first make a distinction between the external mind that allows one to be conscious of the external world, in contrast to the internal "subconscious" mind that actively monitors and regulates the body's internal biological metabolism and related functions.

The external mind is usually capable of doing or thinking about one, or at best, two things at any given time, whereas the internal mind performs tens of thousands of functions every second. This gives one the impression that the internal mind is vastly more sophisticated than the external mind. Most people acknowledge that they have a subconscious mind that is an active partner in their day-to-day activities. This is true whether one is talking to one's self -- or just daydreaming. The ability of the mind to generate accurate "pictures" of mental fantasies that could be defined as "visions" is a common, and well-documented human experience. Observe a musician in the middle of a stirring performance and note how the conscious mind seems to give way as the musician appears to go into a trance. It is as if the subconscious mind takes over.

Similar subconscious experiences have been described by numerous scientists, business executives, entertainers, and others. One notable example was recorded by August Kekule, a German chemist who, in 1890, described how some of his most significant insights on molecular theory came to him in his book, *KeKule's Dream:*

> *"During my stay in London I spent my evenings with my friend Hugo Muller. We talked of many things but most often of our beloved chemistry. One fine summer evening I was returning by the last bus, 'outside' as usual, through the deserted streets of the city. I fell into a dream, and lo, the atoms were gamboling before my eyes! Wherever, hitherto, these diminutive beings had appeared to me, they had always been in motion; but up to that time I had never been able to discern the nature of their motion. Now, however, I saw how, frequently, two smaller atoms united to form a pair; how a larger one embraced the two smaller ones; how still larger ones kept hold of three or even four of the smaller; whilst the whole kept whirling in a giddy dance. I saw how the larger ones formed a chain, dragging the smaller ones after them but at the ends of the chain.... The cry of the conductor 'Clapham Road,' awakened me from my dreaming; but I spend a part of the night in putting on paper at least sketches of these dream forms."*

This subconscious insight was the origin of the "Structural Theory." Other examples describe where someone will spend hours working on a problem without success. Then, perhaps several hours or days later, while that person was relaxing or involved in some other activity, the subconscious mind, which had presumably been working on the problem, suddenly announces the solution to the problem. There are many examples that support the premise that the subconscious mind may, in fact, be a gateway to the microbial or nanobial consciousness that lies within each living organism.

The City of the Cell

Most nanobes live and work within the cells of living organisms. All plants and animals are made up of such cells that are broken down into two major classifications. The smaller and seemingly more primitive cells are called prokaryotes, and these contain no cell nucleus. The more complex eukaryote cells do have a nucleus and are larger than the prokaryote cells by a factor of 10 in linear measurements, and by a factor of 1,000 in volume (refer to Figure 8.9). Note that the eukaryotic cell on the right has a much more complex network of subcellular structures than the prokaryotic cell on the left. All cells that are more complex than bacteria are classified as eukaryotes.

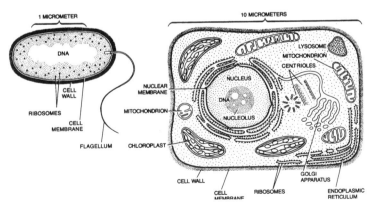

Figure 8.9: Primary Cell Types
From "Archaebacteria," by Carl R. Woese. Copyright
©June 1981 by *Scientific American*, Inc. All rights reserved.

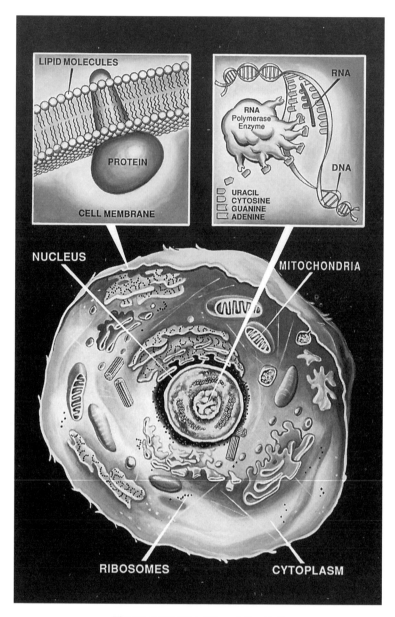

Figure 8.10: The City of the Cell
Biological membranes are made up of proteins, which are the large
structures set in a bi-molecular film of lipid molecules. DNA unwinds
with the aid of an enzyme in order to duplicate its protein codes.

Observing the surface of a cell or other biological structure through a microscope is somewhat like looking at the external surface and structure of a skyscraper or space shuttle. This is because the surface membranes of cells conceal an incredible level of sophistication that exists at the molecular level. In the case of humans, each person is made up of about 100 trillion eukaryotic (i.e., nucleated) cells, and each eukaryotic cell can be viewed as a vast city populated by tens of thousands of microbial citizens. To begin to appreciate the complexity of cell biology, Irving Block, writing in the October 1983 issue of *Science Digest,* described what would happen if one could reduce one's size to that of an oxygen molecule:

> *".... We would find ourselves knocking on the gates of a vast walled city with elaborate defenses and a complex economy. At the gate -- the cell has gates! -- we would be carefully frisked for harmful or friendly intent. Once inside the bristling walls, we would enter a bewildering network of canals and roadways, teeming with the transport of raw goods, finished products and communications. We would see that this traffic knits together a system of warehouses, assembly plants, power stations, and processing plants for rendering raw materials and disposing of waste. All roads, we would find, lead to a moated and walled Kremlin from which issues a constant stream of messengers bearing blueprints and coded orders to all corners of this bustling miniworld."*

In a molecular description of the eukaryotic cell, the nanobial protein manufacturing and assembly plants are referred to as the ribosomes; the power stations are mitochondria, the messengers are other specialized proteins, and the moated "Kremlin" is the nucleus of the cell. Inside the nucleus, the master DNA biocomputer determines what nanobial proteins shall be produced. This, in turn, determines the structure and function of the individual cells; whether they are to become brain cells, bone cells, or any of the other specialized cells that make up living organisms. DNA was first described in 1953 by two investigators, James Watson and Francis Crick. As incredible as it may seem, each cell nucleus contains many feet of DNA, which means each person contains billions of miles of it. The structure of DNA is similar to a ladder twisted like a spiral staircase. Each step in the ladder is made up of four different nucleotides (refer to Figures 8.2). The DNA nucleotides are grouped into sets of three steps each, which contain the codes for the twenty different kinds of amino acids that are the building blocks of protein molecules. DNA is critical because it contains the computer code, or blueprint, that directs the cellular ribosome factories to manufacture an almost infinite variety of proteins. Indeed, if DNA represents the master computer in the heart of the cell nucleus, the computer operators are the nanobe-scale proteins, which might be thought of as the "people" who run the molecular machines in the city of the cell. The molecular structure of the DNA in humans is now being undertaken in what is referred to as the Human Genome Project.

The Human Genome Project

One of the important first steps toward the era of designer genes and molecular medicine will involve mapping the human genome, which is made up of about 100,000 genes, molecule by molecule. To accomplish this undertaking, in 1987 the U.S. Congress authorized roughly $3 billion in funding for the Human Genome Project, which is the most expensive and ambitious undertaking in biology. It is biology's "moon shot" program. And what it will provide will be no less than the chemical blueprint of human life; the code of codes; the Holy Grail for *Homo sapiens.*

The objective of the project is to characterize the operating instructions for the human body, and in the process, it will revolutionize the detection, prevention and treatment of everything from the common cold, to the aging process itself. To date, over $250 million has already been spent, and the project is on budget and ahead of schedule. The task of carefully mapping and describing the construction and detailed structure of the amino acid chains that make up a gene's proteins and nucleotides is a formidable one. Nucleotides are like letters in the genetic alphabet, and when they are linked together, they form the familiar double helix. A typical human genome contains roughly three billion nucleotides, and if a single letter were assigned to each nucleotide, it would take thousands of volumes to record the information. And it is far more likely that it will take many paragraphs, and not a single letter, to record the information contained in a single nucleotide. The initial phase of mapping and sequencing the entire human genome was officially completed on June 27, 2000, but putting all the DNA components in order is only the first step in the effort to compile a complete operating manual of the human body.

According to Dr. George H. Sack Jr. of the Johns Hopkins University School of Medicine, it is already clear that most people are silent carriers of defective genes that will eventually result in major medical disorders. Because of the Human Genome Project, however, it will soon be possible to identify and replace the defective genes with normal ones. But decoding the book of life poses daunting moral dilemmas, and many people will understandably be concerned. With knowledge of our genetic code will come the power to re-engineer the human species. A second major area of research that needs to be intensified is the identification and description of the body's vast population of protein-based enzymes, which play such a critical role in biochemistry and metabolism. Up until now, only a small fraction of the human body's enzymes have been identified or classified, although the fields of enzymology and enzyme engineering are making remarkable progress.

As Eric Drexler points out, proteins and other biochemical molecules can only function in a wet environment, and they are highly sensitive to both temperature and pressure. The "second generation" nanotechnologies that Drexler envisions will not have these limitations. In the same way that relatively fragile human hands were able to construct much more durable machines of steel and plastic, Drexler believes that soon molecular biologists will be able to use existing proteins to make protein-like nanomachines with materials that will enable them to operate in dry environments and withstand both high temperatures and pressures. These nanomachines could also assemble carbon atoms by bonding them, layer by layer, to form a fine flexible diamond fiber that would have over fifty times the strength of the same weight of aluminum. Drexler refers to these second generation nanomachines as assemblers. But he also describes two other important classes of nanomachines: disassemblers and nanocomputers. Disassemblers, as the name implies, will be able to take apart a molecule, atom by atom, and with the cooperation of a nanocomputer, record its structure.

Nanocomputers

As Drexler explains, assemblers will allow engineers to reduce dramatically the size and cost of computer circuits and the speed at which they operate. When engineers can assemble molecules atom by atom, they will be able to build biochip computer circuits in three dimensions, or they may be able to bypass conventional electronic circuits altogether. This is because biochip nanocomputers may be mechanical rather than electrical in structure. Nanocomputers will be able to oversee the molecular work being done by the disassemblers as well as direct the assemblers to make exact copies of the molecule, or modified copies, if that is what is desired.

Drexler makes the point that computation is not dependent on electronic circuits. In fact, the first computers built were not electronic, but mechanical. The reason electronic computers replaced the initial mechanical versions was that the mechanical switches that were connected by rods or wire were large, slow, unreliable and relatively expensive. However, nanomachines could assemble components that would only be a few atoms wide; a mechanical nanocomputer, or nanochip, could be billions of times smaller than the current generation of microchips. Although mechanical signals are about 100,000 times slower than electrical signals, they will only have to travel 1/1,000,000 as far. This means mechanical nanocomputers should be able to transmit signals thousands of times faster than the most advanced electronic microcomputers currently in use. Taken together, the power of nanotechnology assemblers, disassemblers, and computers is virtually limitless.

As Drexler writes:

> *"Molecular assemblers will bring a revolution without parallel since the development of ribosomes, the primitive assemblers in the cell. The resulting nanotechnology can help life spread beyond Earth -- a step without parallel since life spread beyond the seas. It can help mind emerge in machines -- a step without parallel since mind emerged in primates. And it can let our minds renew and remake our bodies -- a step without any parallel at all."*

While there are many positive developments that can be brought by nanotechnology, Drexler goes to great lengths to point out the potential dangers of such developments -- not the least of which are the implications for weapons development. However, one of the most serious concerns involves the power of exponential growth. As Drexler explains, just as cells reproduce themselves, nanomachines will also need to replicate themselves:

> *"Imagine such a replicator floating in a bottle of chemicals, making copies of itself. It builds one copy in one thousand seconds, thirty-six in ten hours. In a week, it stacks up enough copies to fill the volume of a human cell. In a century, it stacks up enough to make a respectable speck. If this were all that replicators could do, we could perhaps ignore them in safety. Each copy, though, will build yet more copies. Thus the first replicator assembles a copy in one thousand seconds, the two replicators then build two more in the next thousand seconds, the four build another four, and the eight build another eight. At the end of ten hours, there are not thirty-six new replicators, but over 68 billion. In less than a day, they would weigh a ton; in less than two days, they would outweigh the Earth; in another four hours, they would exceed the mass of the Sun and all the planets combined -- if the bottle of chemicals hadn't run dry long before."*

There is no more impressive example of how significant exponential growth can be. It demonstrates the power and the promise of nanotechnologies, which can provide unprecedented problems or unprecedented solutions -- depending on how they are directed. As a result, there will be those who would attempt to stop such technology developments. While it is not realistically possible -- or desirable -- to prevent the developments in computers and biotechnology, it is both possible and desirable to provide a reasonable amount of guidance. As with most things in life, it is a question of balance. The fact that assemblers can replicate themselves in such vast numbers means if they are improperly used, they could eliminate life on the Earth. On the other hand, if they are properly used, they could realistically solve the global problems of chemical pollution, resource depletion, and repair the defective genes and molecules that result in the process of aging and limited life spans. Given current funding levels, biocybernetic capabilities will probably not significantly impact health care systems until the year 2030 or 2040. If the amount of funding increased, it would certainly reduce the time required to implement such systems. There is no question, however, that significant progress is already being made. In addition to the biochip developments already mentioned, other examples that demonstrate how rapidly the progress toward nanotechnology systems is occurring include microtechnology devices, the scanning tunneling microscope (STM), and resonance-ionization spectroscopy (RIS) systems.

Microtechnology

Nanotechnology operates on the nanometer-scale, which is one-billionth of a meter. Microtechnology operates on a micrometer scale, which is one-millionth of a meter. A single atom is typically between 1 and 5 angstroms in diameter, and an angstrom is one-tenth the length of a nanometer. The initial research into micromechanisms began in the 1960s in research centers at IBM and various universities, including the University of Michigan, Stanford University, the University of California, the University of Tokyo, and Matsushita Research Laboratories in Osaka, Japan. The first peer-review paper on microtechnologies was published in May of 1982 by the Institute of Electrical and Electronics Engineers. The paper was prepared by Kurt Petersen, who at the time was with IBM's research division in California. Petersen is now executive vice president of NovaSensor, a Fremont, California, company devoted to microstructures. One of the first prototype microtechnology devices was developed two years later, in 1984, by Microsensor Technology Inc., which is also located in Fremont.

The system was a micro-monitor Universal Gas Analyzer, which could identify up to 100 different gases at the rate of 10 gases every 45 seconds. The progress in microtechnologies has increased dramatically. This is demonstrated by the fact that numerous major corporations in Japan and the U.S., including IBM, Allied Signal, AT&T, Honeywell, General Motors, Ford, and

Figure 8.11: Micromachined Gears
The image was reprinted with permission from *Popular Science,* Times Mirror Magazine, 1989.

Chrysler, are now committing hundreds of millions of dollars to microtechnology developments. Many engineers in the field believe that microtechnology systems will be as revolutionary as the integrated circuit. Much of the initial work in microtechnology was undertaken in the area of sensors. Starting in 1987, however, a whole new category of microtechnologies became available called "microactuators." Microactuators are micron-scaled pin joints, springs, cranks, sliders, and gears (refer to Figure 8.11). Thus far, the primary material used in microtechnology systems has been silicon, which is comparable to steel in strength. Yet, because silicon is a single-crystal material, it can flex without fatigue almost indefinitely. Silicon also has piezoresistive characteristics. This means that, under strain, silicon's electrical resistance changes, which is especially useful for sensors applications. At AT&T laboratories in Holmdel, New Jersey, investigators have even manufactured micron-scale turbines that rotate with speeds in excess of 24,000 revolutions per minute (rpm).

An electron microscope shows silicon gears in Figure 8.11 with teeth that are only 15 microns across, which is less than one-fifth that of a human hair. Investigators at Stanford University's Department of Electrical Engineering have already developed a micro machined electric-to-fluidic actuator valve that can be integrated directly into a computer chip. This is a key component that will be required in the development of advanced robotic and cybernetic systems. The advantages of miniaturized micro mechanical parts with rotatable joints or sliding members are numerous; they have a much higher speed capability, they are considerably more accurate, they are less sensitive to temperature extremes, and most importantly, they will be substantially less expensive to produce. Because of these factors, the potential multi-billion dollar micro technology market is simply staggering, particularly in automotive, aerospace and biotechnology applications. Recognizing this, the National Science Foundation has called for a $50 million national laboratory for micro dynamics development.

Investigators in Japan are already developing a micro technology robot that is expected to be less than four-hundredths of an inch in diameter. Such a micro robot will be able to travel through veins and inside organs. Iwao Fujimasa of Tokyo University's Research Center for Advanced Science and Technology believes their micro robot will be able to transmit its location and findings and otherwise treat diseased tissue. The micro robot project is made up of scientists from several universities as well as companies such as Toyota, Hitachi, and Sumitomo Electric Industries. The highly influential Ministry of International Trade and Industry (MITI) has also provided financial support for Fujimasa's micro technology project.

Nanoscopes

As impressive as microtechnologies are, one of the more dramatic examples of how rapidly the nanotechnologies are developing is the scanning tunneling microscope (STM). The STM was first developed in 1981 by two IBM scientists, Heinrich Rohrer and Gerd Binnig. Five years later, they would share the 1986 Nobel Prize in Physics for their efforts with Ernst Ruska of the Haber Institute in West Berlin for his work in electron optics and his design of the first electron microscope in the 1930s. It was highly unusual for the Royal Academy of Sciences to award the Nobel Prize so soon after the initial research was concluded. But as the Royal Academy stated, the electron microscope was considered one of the most important inventions of this century, and the STM was as much as 100 times more powerful than the world's best electron microscope. The STM is capable of imaging the actual surface of individual atoms that are, in turn, the structural components of molecules. The STM has no lens. Rather, it has a stylus with a single atom at its tip that tracks over a specimen much like a needle follows the grove on a phonographic record. Although the STM stylus normally does not actually touch the specimen while scanning, it certainly has the capability to do so. R. Fabian Pease, a professor of electrical engineering at Stanford University, has indicated that the stylus tip of the STM can be used to print fine lines on integrated circuits that are less than one-half micron (i.e., less than 500 nanometers).

Such precision in a cutting edge will be highly useful in nanomachining, material deposition and removal, and in a variety of chemical and lithographic processing procedures. This means the STM is one of the first instruments that has been developed with true nanotechnology capabilities. In addition to the STM being much more powerful than an electron microscope, it can also examine objects in their natural state, without damaging the specimen. STMs are also substantially less expensive than electron microscopes. They range in price from $30,000 to $100,000, in contrast to electron microscopes that cost around $800,000. As a result, they are spreading rapidly in research laboratories around the world.

As advanced as the STM technology is, it has resulted in an even more sophisticated technology that has been developed by investigators at IBM and the Department of Physics at the University of California at Santa Barbara. This even more advanced nanotechnology is referred to as the "Atomic Force Microscope" (AFM). The AFM and STM instruments are similar, in that they both utilize a probe that terminates with a single atom to track a specific molecular surface. However, unlike the STM, the AFM comes in actual contact with the material it is scanning, and senses the minute forces that exist between the atoms. Both the STM and AFM instruments are capable of atomic manipulation as well as resolution.

The primary advantage of the AFM is that it is not restricted to tracking only conductive or semi conductive surfaces, as is the STM instrument. This means the AFM has the potential to be more easily used in aqueous environments and with a much wider range of organic and inorganic elements and molecules.

Resonance-Ionization Spectroscopy

Another promising nanotechnology, referred to as resonance-ionization spectroscopy (RIS), is being developed utilizing
finely tuned lasers to isolate and manipulate individual atoms.
Writing in *Scientific American,* Vladilen S. Letokhov, who is head
of the Department of Laser Spectroscopy and deputy director of
the Institute of Spectroscopy of the Russian Academy of
Sciences, described his investigative assumptions:

> *"The light spectrum is the language 'spoken' by
> atoms and molecules. Just as different peoples
> speak different languages, so each kind of atom
> or molecule emits and absorbs light of charac-
> teristic wavelengths, or colors. By measuring
> these wavelengths, investigators over the past
> 100 years have made great strides in under-
> standing the properties of all the elements in the
> periodic table and of an enormous number of
> their molecular compounds."*

Letokhov is not alone in his investigation of RIS systems. G.
Samuel Hurst, who works at the U.S. Oak Ridge National
Laboratory, has also developed an RIS system that utilizes highly
sophisticated, ultra sensitive laser analysis technology. RIS systems are so sensitive they are capable of being tuned to a specific wavelength of a particular atom or molecule. As a result,
they can locate an individual atom in a given mixture, ionize it (by
stripping away one of the outermost electrons) and then separate
it by applying an electrical field.

The fact that RIS systems can locate individual atoms and
then isolate them provides a dramatic new capability for investigators to explore and measure the structure of the various
species of atomic nuclei. The RIS technique is extremely efficient; its operation time is determined by the duration of the laser
pulses. In typical experiments the pulses are about 10 nanosec-
onds (one billionth of a second) long. In the near future, operation times are expected to be reduced to the range of picoseconds (trillionths of a second) or even femtoseconds (thousandths
of picoseconds). With the development of the Scanning
Tunneling Microscope, the Atomic Force Microscope, and

Resonance-Ionization Spectroscopy systems, it is clear that the nanotechnology era has already begun. This only underscores how rapidly technology, in general, and biotechnology, in particular, is developing. In the final analysis, such technologies provide great hope for the future, for they represent the beginning of the era of molecular medicine that will result in a designer gene era that will, in the process, provide indefinite life spans.

Overpopulation Considerations

Of all of the problems that currently confront the human community, one of the most significant is clearly the fact that there are more and more people who are competing for fewer and fewer resources. Many of the most serious environmental problems are also directly related to human population increases. Due to advances in food production and medicine, the average human life span has already been increased from about 40 years to over 70 years, an increase of more than 70 percent. As life spans are progressively increased due to the rapidly approaching era of molecular medicine, many people are understandably concerned about the obvious problems associated with overpopulation.

However, with the ability to accurately regulate the genetic code and the related protein molecules that determines one's physical being, it will be a relatively simple matter to switch off the specific genes that are responsible for reproduction. This means the same genetic formula that would allow individual cells continually to regenerate themselves could also contain the specific gene sequence that would prevent fertility. With the technical capability to switch-off reproductive genes, the emotional issue of abortion will finally be fundamentally resolved. More importantly, however, it means an element of birth control could be exercised for those who choose to regenerate their cells and organs. Such measures would only apply in cases where the carrying capacity of the environmental life support system could not effectively accommodate the increasing numbers of people.

The immediate concern of overpopulation will only be a serious problem as long as humans are limited to living on the surface of the Earth. Given the diminishing resources that are already occurring on the Earth, it is reasonable to assume that if humanity survives the oblivion scenario, large numbers of people will be living in large space habitats that could look similar to Starship Hydrogen discussed in Chapter 5. These human engineered planets would be designed to rotate in order to simulate the Earth's gravity. This being the case, humans and other mammals could remain in space for extended periods of time without having their bodies turning into a shapeless, jellyfish type of organism. On the other hand, by the time space colonies are a reality, biocybernetic organisms will be also, and they will probably not require gravity to maintain their shape because their molecular bonds will be engineered to withstand zero gravity environments. In any case, due to the lack of gravity, it should be very inexpensive to build large superstructures in space, provided enough of an infrastructure exists to mine the necessary resources from the asteroids.

This implies that with the relatively unlimited resources that exist in space, there would be no need to place a limit on human population levels, regardless of how long people might choose to live. Drexler points out, however, that if a molecular-scale replicator begins to exponentially copy itself every one thousand seconds, at the end of ten hours there would be over 68 billion replicators; in less than two days these replicators would outweigh the Earth, and in another four hours they would exceed the mass of the Sun and all the planets combined. This underscores the reality that unlimited growth cannot be maintained indefinitely, even in the apparent vastness of space.

While it is clear that the development of large-scale space colonies is a logical, natural -- and indeed -- a necessary objective (much like a chicken coming out of its planetary egg), this does not mean to suggest that the exponential growth of the human population could continue forever. It does, however, suggest that there will be ample time to consider such a truly long-term problem if the more immediate energy and environmental problems facing the global human community are resolved. Perhaps the greatest tragedy of all is knowing that although such a biocybernetic-immortalist era is not very far away (perhaps 30 to 50 years into the future), there will be many people who will die before the major advances are made. But there is a parachute, of sorts, that is available to most people now living. It is referred to as cryogenic suspension, or biostasis, and it involves using ultra-cold temperatures to preserve biomolecular structures.

The Cryogenic Ark

Although what is currently defined as death may come at any moment for any number of reasons, in most cases, it is possible to preserve the biochemical molecules that reside in the brain (neuropreservation) with cryogenic (i.e., low-temperature) suspension. This "suspended animation" can be maintained virtually indefinitely if liquid nitrogen temperatures are used. Thus, if one has preserved the molecules that make up the neurons in brain tissue (which have been encoded with an individual's memories and personality), one has, in effect, preserved the fundamental essence of that person. While the thought of just preserving one's head and brain, instead of the rest of one's body, is psychologically difficult for some people to accept, it is important to be aware that each person started out as a single cell, and that the genetic information contained in that cell, along with most other cells in the body, has the code for generating all of the body's organs and support tissue. However, the DNA does not contain the molecules of memory, which are the only significant molecules to preserve.

Although the biotechnology to revive and regenerate successfully a cryogenically preserved brain, along with the rest of the body's organs, is not presently available, such procedures could be commonplace by the year 2050. It is not a question of where medical technology is today, but where it will be in the foreseeable future. Given the advances in cryobiology, molecular biology and nanotechnology, the concept of being "born again" is going to take on a new meaning. The first serious book on the subject of the cryogenic suspension option and its obvious and profound sociological implications was *The Prospect of Immortality* (first published in 1962), authored by Robert C. W. Ettinger, a physics and mathematics instructor at Wayne State University in Detroit, Michigan. At present, cryogenic suspension it is the only option available that can preserve the molecular structure of the brain in the event that one's "death" should occur before the Biocybernetic Era of molecular medicine arrives. An excellent assessment of the cryogenic option was provided by James B. Lewis, a chemist and molecular biologist who received his Ph.D. in chemistry from Harvard University. For more than a decade, Lewis has been a principal investigator in research projects funded by the National Science Foundation, the National

Cancer Institute, and the American Cancer Society. Lewis is also an affiliate associate professor in the Pathology Department of the University of Washington Medical School. When asked to provide an affidavit of expert testimony on the cryogenic option, Lewis responded with a five-page document, which is excerpted as follows:

> *"I am writing to provide a professional opinion as to the general feasibility of cryogenic preservation as a rational gamble to obtain a chance of the future reanimation of the very recently deceased. These comments are based on the reasonable extrapolations as to the capabilities of future technologies, founded upon our current knowledge of basic molecular and cellular biology. The conclusion of my arguments will be that, although reanimation of cryogenically preserved individuals is not possible with current technology, nor can it be proven that reanimation will be possible in the future, current developments most emphatically do provide a reasonable basis for supposing that such technology might be available within the next 30 to 50 years.*
>
> *Further, and most importantly, the nature of these probable developments is such that damage caused by current freezing techniques (which is serious enough that reanimation with current techniques is highly unlikely), is not likely to prevent reanimation with radically different future techniques. Thus the decision to pursue cryogenic preservation now is a rational gamble and is definitely not absurd or without basis in our current scientific knowledge."*

Lewis makes a key point about the fundamental distinction between the preservation of biological function and preservation of structure. Lewis has carefully read Drexler's book *Engines of Creation* and is convinced that nanotechnology is a reasonable technological expectation. He has even formed a technical study group in the Seattle area to explore the development and implications of nanotechnology. As a result, he argues that if the technology to manipulate complex molecular structures on the atomic level becomes available:

"....it will be possible to repair freezing-induced damage, and thus to rebuild and then reanimate. The relevant question then becomes what degree of structure, not function, must be preserved to preserve the essential aspects of an individual's personality. . . the very technology that will make it possible to repair freezing damage to the brain will make it possible to rebuild a complete body, from the genetic code preserved in each cell of the individual from scratch."

Defining Death

Medical technology is already far ahead of the legal system. People who are pronounced dead essentially have no rights. But the question is: *What is the correct definition of death?* Presently, death is usually pronounced if a person's heart and/or breathing cease for more than five minutes, or if there is no measurable brainwave activity. Under this definition, however, there are many people who have technically "died" and have been successfully brought back to life.

One notable example of this new medical technology is provided by Dr. Robert Spetzler and his colleagues who work at the Barrows Neurological Institute in Phoenix, Arizona. Spetzler is a specialist in operating on aneurysms, which are a local enlargement of a weakened wall of an artery that is typically caused by a disease or injury. Because the aneurysm is essentially blown up like a balloon, if it is cut, a massive brain hemorrhage can often result, and the patient usually does not survive. In order to overcome this problem, Spetzler and his surgical team have successfully developed and utilized a surgical procedure that involves chilling a patient's body temperature to about 15 degrees C (59 degrees F). The operating room is also cooled to the same temperature during the brain operation. This procedure has now been successfully completed on dozens of people, whereby all of the patient's blood is removed, which allows the aneurysm to collapse so that the surgeons can then remove the weakened tissue. But during this process, the patient's heart and breathing also cease, and the brain cells stop exchanging electrical signals for nearly an hour. Such symptoms would be classic definitions of legally pronouncing a person to be dead.

Figure 9.12: Ethel St. Lawrence
Ethel. St. Lawrence had to die in order to to be saved. During a brain operation, doctors had to lower her body temperature to 15 degrees C, they then stopped her heart, all of her blood was then removed and all of the electrical activity of her brain ceased for more than 40 minutes.

There was also a widely publicized case of a child who had fallen through ice in the middle of winter and remained under water and unconsciousness for over an hour. Most people would have considered the child to be dead, yet because of the circumstances of having the icy water slow the child's metabolism, he was successfully revived. These examples underscore that fact that the current definition of death is no longer adequate given the rapidly changing state of medical technology. It would seem reasonable that death should be defined as a condition whereby it is impossible to restore one to life. Specifically, a more accurate definition of death would describe a condition where the biochemical molecules of memory that are stored in the brain are irreparably damaged or lost (such as in a fire or explosion), for only in such a condition is a person truly beyond recovery, and hence, truly and permanently dead. Given this definition, if a person's molecules of memory are cryogenically preserved, then the person is not dead. Rather, they are in a state of suspended animation. Given the exponential advances that are now occurring in the field of molecular medicine, it is only a question of time before such individuals will be successfully revived.

The Ultimate Disease

If youthful life spans had no chance of being dramatically extended, the cryogenic option would not be so appealing. How and why cells age has been under study by numerous investigators for many years. In 1962, Bernard Strehler, a professor of biology at the University of California, authored one of the first books on understanding the biological mechanisms of aging: *Time, Cells & Aging.* Strehler believes that aging is essentially a molecular-based disease, and as such, it would lend itself to reductionist scientific analysis. Thus, eventually a cure is going to be found. It is only a question of how soon.

A decade ago, a common theory on aging held that early in the life of a single cell, the DNA molecule was new and its coded instructions that result in the production of proteins would be clear and accurate. But it was presumed that with time, errors would accumulate in the code, and eventually, the cells would become defective, diseased, or they would die altogether. When enough cells in an organ would die or become diseased, the organ would fail to function and the organism would die. However, given the many developments in molecular biology in recent years, it is now the opinion of many investigators that aging and death is no accident. They believe mechanisms of aging are deliberately programmed into the DNA code. The reasoning is that after the organism produces an offspring, its primary biological function has been fulfilled. It is during this process of bisexual reproduction, that a natural process of "genetic engineering" occurs as the DNA code is modified with each new organism. In this way, biological organisms are continuously able to adapt to an ever-changing environment.

Until recently, the complexity of the aging process seemed too daunting to dissect. But according to an article "Few Dozen Genes Rule Old Age" published in the March 31st issue of the *Washington Post,* that is about to change. In a new report published in the Journal *Science,* scientists have been able to accurately track gene activity over the course of a lifetime with what are referred to as "gene chips," which are DNA-coated microchips. The gene chips are able to trace the activity of tens of thousands of genes individually over time with high-powered computers that can detect specific patterns in what would otherwise be biochemical chaos.

As a result of this powerful new tool of molecular biology, investigators are now able to begin understanding how specific genes influence the mysterious and inevitable process of aging. According to Richard Lerner, President of the Scripps Research Institute in La Jolla, California, and a lead scientist on the study:

> *"We have spend 100 years of reductionism trying to understand biological processes by looking at the individual pieces. Now we are going to try to understand complexity, which is how the pieces work together."*

Lerner and his colleagues took live skin cells from two youngsters who were ages 7 and 9, two 37-year olds, three people in their 90s and three children with a rare disease called progeria, which makes children age prematurely, causing the children to die in their teens from heart attacks or other diseases of old age. The team compared the activity levels of more than 6,000 genes among the various age groups. Most of the genes maintained steady activity levels in youth, middle age and old age. But 61 genes "speeded up" or "slowed down" by a least twofold as they made the transition from one age group to the next. Of those 61 genes, fully one-quarter are what Lerner calls "quality control" genes, which are supposed to prevent cells from dividing if those cells have accumulated significant age-related chromosomal defects.

Other theories on aging make reference to the damage that occurs at the molecular level, and there appears to be a direct relationship with reduced enzyme activity and the aging process. If this is the case, when nanotechnology and designer gene systems become available, it will be possible to correct such molecular damage and allow the enzymes, and other key biochemical molecules to remain in optimum operating condition. In effect, the human body (or any other biological organism) will have then made a biological transition to renewable resources. Given the current knowledge and information explosion, the human brain has been forced to evolve primarily from a sensory reaction organ to a highly specialized information processing and storage system. It can now take 20 to 30 years to master even one discipline, and when a new child is born, the entire educational process must start all over again.

While bisexual reproduction and limited life spans has been the nanobial method of genetic engineering in the past, in today's world, the loss of highly experienced and trained individuals is difficult if not impossible to calculate. This has always been true to some extent, but for the first time in human history, it is possible to do something about it. Specifically, it will involve switching off the biological mechanisms of aging by reprogramming the necessary enzymes and specific genes that regulate the aging process. Essentially, molecules would be continually be regenerated as needed, thus tissue and organs would not be subjected to the ravages of aging.

Those who are skeptical about the future ability of molecular medicine to arrest the biological process of aging, or repair and regenerate a cryogenically preserved individual are those who are, typically, not very familiar with the recent developments in molecular biology or biotechnology -- not to mention protein engineering or nanotechnology. The attitude of such people is similar to when some people thought only birds should fly, or that it would never be possible to place a man on the moon. Even today, there are some people that refuse conventional medical treatment for themselves or their families because they view such progress as "unnatural."

Ethical Questions

There is no question that the implications of research into molecular biology, microtechnology, and nanotechnology are both profound and unprecedented. There will, of course, be a certain number of people who will simply not support such medical advances. Ultimately, each person must determine the value of his or her life. But while it is natural that some people will be fearful of such far-reaching developments, it is reasonable that most people will conclude the potential benefits will more than offset the risks involved. Disease and death have always been the ultimate fear of most people, and if it becomes possible to delay such unfortunate realities through molecular medicine, few people would likely refuse such treatment. *How many people would allow a tooth cavity to go untreated?* The bottom line is that most people will take all reasonable steps to avoid death or a serious illness.

But a deeper ethical question remains: *Is providing an unlimited life span for individuals a worthy objective? Or should such developments be banned as dangerous and unnecessary?* It should be clear that the explosive advances in biotechnology have already far outpaced the traditional ethical, legal and medical definitions of life and death. Philip Lipetz, a molecular biologist at Ohio State University has stated that, "discoveries are tumbling over breakthroughs in quick succession, and scientists are coming very close to manipulating life spans." He has expressed a concern, however, that, although scientists are trained to understand the biological and molecular mechanisms of aging and to figure out ways to manipulate them -- these scientists are not trained to handle the substantial emotional impact that these discoveries are going to have on society. Indeed, it is the potential emotional backlash that keeps many scholarly investigators from making public comments about where the technology is heading.

Having an unlimited life span does not necessarily imply immortality. Sooner or later something could certainly go wrong. Death can surely be delayed, but how long it can be avoided is unknown. A biocybernetic individual, however, would be very difficult to destroy without his or her -- or its -- permission, because each and every atom of a biocybernetic person will be "backed up." Thus, even if biocybernetic person were vaporized in a nuclear explosion, he or she could still be completely regenerated. In any case, there is a significant difference between living 50 or 70 years compared to having an opportunity to live for thousands, or perhaps millions of years. With such life spans, who is to say what is going to be possible. From our current perspective of time, we can only begin to appreciate the possible life one could have as an essentially immortal biocybernetic being. But it is a journey that is conceivably available to anyone now living. And that is the important point to understand.

The subject of death is both awesome and universal. Although it affects everyone, most people don't like to think about it, but the reality is that eventually something is going to bring us down. It is the ultimate event for each conscious being. The difference now is that for the first time in history, it is possible to do something about it. This, of course, means that new philosophical outlooks will evolve to provide the appropriate ethical and moral perspectives that will logically follow from such changes.

From the perspective of the people involved in the cryogenic organizations, there is no question that the only thing worse than dying and being frozen is dying and not being frozen. As the pop artist and writer Ashleigh Brilliant has stated so concisely in one of his "Pot Shot" cards:

> *"I'm prepared for anything*
> *but is death anything?*

Woody Allen made a similar concise thought in the script from the movie *Sleeper:*

> *"I don't want to achieve immortality through my*
> *works or my descendants. I want to achieve*
> *immortality by not dying."*

Priorities

A though the cold war with the former Soviet Union is over, the U.S. government continues to spend billions of dollars annually on new and improved weapons systems. Indeed, over $70 billion was authorized by the Congress to build the new stealth weapons delivery system. Yet, not even $1 billion is being spent on research into nanotechnology in general -- much less in the specific areas of understanding the biological mechanisms of aging. Imagine how rapid the progress of curing disease would accelerate if the U.S. reallocated its spending priorities. Of course, there will always be some people who do not seek to have a longer productive life. Many teenagers take their own lives in spite of the fact that their whole life is ahead of them. It would, however, be interesting to know how many suicidal individuals would change their mind and their emotional outlook on life if they knew that they had an immortalist biocybernetic future ahead of them. There are also many people whose religious beliefs dictate that death is necessary to achieve everlasting life. It is interesting to note that they are still seeking immortality; they just feel that they have to die in order to obtain it.

Although this is a personal matter, it is worth noting that the death of a loved one is a deeply felt tragic event. As a result, it is psychologically reassuring to believe that the essence of the person (i.e., the person's soul) will live on forever in Heaven. The problem with this view from a hard science perspective is that one can logically conclude that when the body ceases to function, the eyes will no longer be able to see, the ears will not be able to hear, the voice box will not be available for speech, and the brain that contains our memories, our emotions and sum total of the molecules that make up our individual personalities will have decomposed. This means the senses, which are the instruments of consciousness, will have been eliminated. Thus, if there is a Heaven, it is highly unlikely that our human consciousness could have any meaning in such an environment.

Summary

There will be many profound issues and questions to consider as a result of the changes that are being brought about by the extraordinary advances that are occurring in the interrelated fields of chemistry, computers and biotechnology. The very structure of contemporary human society and culture is based on a 70- or 80-year life span; and if and when life spans become indefinite, everything is going to change in ways that cannot be imagined at present. Humanity is on the threshold of literally evolving not only into a new species, but into a new life form. *Homo sapiens* is rapidly evolving into *"Homo Immortalis."*

What is especially interesting is that these remarkable developments in biotechnology are virtually inevitable unless a systems collapse is brought about by some catastrophic event, such as a nuclear war or the contamination and/or disruption of the Earth's biological life-support systems. As such, the coming age of designer genes is not so much a question of whether -- but when. If the U.S. government would provide the same level of funding for molecular biology as it does for weapons systems, the progress in biotechnology could be greatly accelerated. Alan Harrington, author of *The Immortalist* (published in 1969 by Random House), took the view that given the new capabilities of molecular biology and biotechnology, "We should hunt death down like an outlaw."

If it is possible to arrest and reverse the biological process of aging and eliminate disease as we now know it, it is hard to imagine what priority could possibly be more important.

It is especially significant that anyone now living has the possibility of evolving into a biocybernetic organism, and thereby participate in the "Immortalist Era" that is rapidly approaching. Cryogenic suspension may become necessary, but there is no question that it is theoretically possible for anyone now living to take part in the most profound evolutionary change since the origin of life itself. The implications of such developments are so far-reaching, they are hard to put into perspective. The one consideration that cannot be dismissed is that the surface of the Earth may no longer be habitable by the year 2050 due to the exponential decreases in the Earth's stratospheric ozone shield. As such, biocybernetic organisms may be the only life forms capable of adapting to such a hostile environment.

The Earth is ultimately a doomed planet in any case; it is only a question of time before our Sun is going to exhaust its supply of hydrogen fuel, and during the Sun's final stages of its life, it will expand and engulf the Earth in its decaying stellar fires. This consideration reinforces the reality that we and our descendants are eventually going to have to leave this planetary cradle. In spite of this unfortunate reality, however, it makes no sense to hasten the day when the surface of the Earth will no longer be able to support life. If the exponential destruction of the Earth's biological life-support systems continues, there may not be anyone left to fulfill the dream of an immortalist, biocybernetic utopia.

If the chemical contamination of the Earth is, in fact, unavoidable, we can only hope that in humanity's final days, it will be able to give birth to the ultimate life form: an immortalist biocybernetic species. In the final analysis, the majority of the people who presently occupy what Buckminster Fuller referred to as "Spaceship Earth" need to be aware that they could be developing the technologies of life instead of the technologies of death. It is simply a question of priorities.

Congressional Record

United States
of America   PROCEEDINGS AND DEBATES OF THE 101st CONGRESS, FIRST SESSION

| Vol. 135 | WASHINGTON, FRIDAY, MARCH 17, 1989 | No. 31 |

Senate

HYDROGEN LEGISLATION: A BILL FOR ALL INTERESTS

Mr. MATSUNAGA. Mr. President, I rise to introduce, together with my senior colleague from Hawaii, Senator INOUYE, the junior Senator from Colorado, Senator WIRTH, the senior Senator from Rhode Island, Senator PELL, and the junior Senator from Connecticut, Senator LIEBERMAN, legislation which advances a multitude of causes.

Seldom can a single bill be said to address the gamut of legislative issues, both environmental and economic, facing a new Congress and a new administration. After all, what common thread exists between global climatic change and those activities spawning greenhouse gases, on the one hand, and the U.S. trade deficit and American competitiveness in world markets, on the other?

Energy is the thread, Mr. President, and a national program for hydrogen research and development is the measure that touches all the aforementioned bases.

The form of energy we use is at the heart of virtually all environmental issues. The location of our energy sources is at the crux of our trade imbalance. And in no area of economic endeavor is overseas competition more keenly felt than in the quest for new energy technologies.

For all these reasons I am once more introducing the National Hydrogen Research and Development Program Act, legislation I first offered in the 97th Congress and have urged ever since. Mine has not been a lone voice on this subject, Mr. President. Throughout his career in this Chamber former Senator Dan Evans cosponsored and strongly advocated my hydrogen legislation. In this connection, it is significant that Senator Evans has been the only professional engineer to serve in the Senate in recent memory, save for Mike Mansfield, who was a mining engineer, and the late Stewart Symington, who was self-taught in mechanical and electrical engineering. If they were still with us, I am certain that both would be cosponsors, Mr. President. Moreover, the principal sponsor of companion legislation in the House, Representative GEORGE BROWN of California, is one of the very few scientists serving in that body.

Mr. President, hydrogen is one of the most abundant elements in the universe, with water, a primary source of hydrogen, covering three-fourth of the Earth. Indeed, hydrogen plays a role in such varied, everyday products as peanut butter, vitamin C, and aspirin, not to mention such larger products as clear plate glass windows.

As a transportation fuel, hydrogen's environmental benefits are particularly apparent, as was evident by its inclusion in the national energy policy legislation offered by Senator WIRTH in the last Congress to address the concerns over global warming, acid rain and the greenhouse effect and introduced again this year. Moreover, hydrogen can be transported more efficiently and at less cost than electricity over long distances.

While hydrogen has definite environmental advantages over fossil fuels, because the product of hydrogen combustion with air is essentially water vapor, it also offers benefits in the utilization of numerous energy alternatives—ranging from coal and natural gas, to nuclear as well as to solar and the renewables. Injected into declining natural gas fields, hydrogen can serve as an enhancer, stretching out the life of dwindling supplies.

For those concerned with the interests of the coal industry, hydrogen also figures in an attractive scenario. If coal-gased reactors were to be built at the seashore, they could eject carbon dioxide into the sea instead of into the air, and transmit energy in the form of hydrogen from coal. It is claimed that this could give us perhaps another half century of coal availability without adding anything to the world greenhouse effect.

For those interested in advancing nuclear power, hydrogen can be seen as a vehicle for hurdling the safety barrier. Because energy is cheap to transport long distances with hydrogen as the storage medium and after 300 to 400 miles, increasingly cheaper

Figure 9.1: Hydrogen Legislation

CHAPTER 9

PRIORITIES

"Real truth and insight is never to be found in bits of information or scientific facts, but in the order and the patterns that one chooses to make with them."

Professor Henry Chadwich

A primary thesis of this book is that because of the exponential impact of the global environmental problems and the development of technology in general, humanity is on the threshold of both oblivion and utopia. Due to the complexity of the many environmental factors involved, it is not yet possible to predict accurately which outcome is going to evolve. But because virtually all of the significant global events are accelerating exponentially, there is little question that the transformation time is rapidly approaching. There are plenty of reasons to be concerned. Roughly one-quarter of the human population is already seriously malnourished. That means in exponential terms, it is already 11:58. The stunning advances in computers, artificial intelligence and molecular biology are continuing, but so are the many environmental problems that now threaten the survival of humanity itself. The existing fossil fuel and nuclear energy systems that currently power the exponentially expanding global economy are clearly unsustainable. Moreover, if the environmental costs of these traditional energy sources were factored into their end-use cost, they could never be economically competitive with the renewable energy technologies. At the start of the Twentieth Century, there were only roughly 1.5 billion people on the Earth, and virtually no automobiles.

At present there are more than six billion people and rough-ly a billion automotive vehicles. But unfortunately, no one knows how long the present levels of growth and consumption will be able to continue. Major food production systems are already in serious trouble due to the impending death of the oceans, global chemical contamination, salt accumulation in the soil, droughts and other climatic disruptions that are being intensified by global warming, stratospheric ozone depletion, and the global process of deforestation. The warning signs are everywhere. It is hard to pick up a newspaper or magazine, or watch a television newscast that does not report on some serious global environmental or economic problem that is continuing to degrade. Given these obvious warning signs, it should be clear that the people on *Spaceship Earth* are like the passengers on the *Titanic*. There is no longer any question that we are going to hit the iceberg. The only questions are: *Will there be any survivors; and if so, what is their existence going to be like?*

Solutions

If the human community is to survive the many exponential-ly compounding energy and environmental "icebergs" that lie ahead, it will likely be dependent upon whether the following actions are implemented in time:

1. There needs to be a fundamental industrial transition from the "oil economy" to a renewable "hydrogen econ-omy." Given the exponentially worsening nature of the global environmental and social problems, as well as the exponential consumption of the Earth's remaining fossil fuel reserves, this "transition of substance" needs to occur with a sense of urgency that is typically asso-ciated with war.

2. The industrialized countries should collectively agree that there needs to be an immediate international effort to stop any further deforestation. This effort should also be coupled with a major program of reforestation. The vast forests of the Earth, which are the lungs of the Earth's biosphere, are being exponentially consumed, and this trend must be reversed as soon as possible.

3. Federal tax incentives should be provided to encourage the widespread development of "controlled-environment" indoor food production systems. In the event of a serious disruption of the existing agricultural infrastructures, such modularized "lifeboat" technologies could be the determining factor in whether or not civilization as we know it survives. The basic premise is that as long as people can eat, they will be able to cope with most other dislocations.

Lifeboat Agricultural Systems

Given the multiple environmental and related climatic dislocations that could disrupt the existing agricultural systems, one of the most important safeguards that could be implemented would be the implementation of large numbers of modularized indoor agriculture and aquaculture systems. These automated systems are able to utilize artificial lighting to grow very high-quality produce and seafood. Such systems are like lifeboats because they can operate regardless of weather conditions; they require no conventional farm or fishing equipment; and no toxic pesticides or herbicides are needed. Moreover, a one-acre controlled-environment facility in DeKalb, Illinois, called Phytofarms of America, Inc., has been able to grow as much produce as a 100-acre conventional outdoor farm, without any topsoil, with only one-tenth as much water, in about half the time.

What these lifeboat agricultural systems do require is a reliable source of energy.

The Phytofarm facility has been growing produce commercially for over 15 years. The operation was developed and operated by Noel Davis, a mechanical engineer by training who graduated from the Massachusetts Institute of Technology. The one-acre Phytofarm facility is able to produces about four tons of spinach and herbs weekly for supermarkets and restaurants in several states. Similar success has been achieved in aquaculture with growing high-quality seafood. These indoor agricultural systems are modularized units that could be rapidly expanded if they are needed to replace crop losses that could result from a combination of exponentially compounding environmental problems, that could include stratospheric ozone depletion, salt accu-

Figure 9.2: Controlled-Environment Agriculture
An external view of the Phytofarm controlled-environment agricultural facility in DeKalb, Illinois, which is about the same size as a supermarket. The photograph provided courtesy of Phytofarms of America, Inc.

Figure 9.3: Phytofarm Facility
An indoor view of the one-acre Phytofarm facility that has the same agricultural yield as a 100-acre conventional outdoor farm. The photograph was provided from James Kilkelly Photography, New York, NY.

mulation in the soil, drought, acid deposition, topsoil erosion or the death of the oceans. Indoor agricultural systems could in many cases be integrated into local supermarkets. If food is produced where it is consumed, the transportation costs are eliminated and freshness is maximized. As a result, even if the surface of the Earth became inhabitable, civilization could continue to operate in artificial environments indefinitely. While the thought of living in totally artificial environments may seem unpleasant, it is well to remember that most urban dwellers already spend most of their lives in such environments -- without the aid of lighting systems that simulate the natural outdoor electromagnetic environment.

It will hopefully be possible to avoid the complete destruction of the Earth's biological life-support systems. But it is likely that increasing numbers of the people will not survive the environmental and climatic dislocations that are already underway in much of the world. Even in a "worst case" scenario, however, indoor agricultural systems could allow human civilization to continue -- and indeed to grow and prosper. The basic premise is that as long as people can eat, they will not panic, and thereby survive. The most important question is: *How many of these "lifeboat" systems can be developed before the more conventional food production systems fail?*

This underscores the fact that while there are many formidable long-term problems facing the human community, there are also many technical options available to deal with these problems. It is only a question of priorities. What is significant is that the primary obstacles to implementing such changes are not technical, or even economic; they are educational and political. Unfortunately, the exponential nature of the problems is compounded by the fact that the U.S. Congress is generally preoccupied with legislative efforts that could be characterized as rearranging the deck chairs on the *Titanic*. It brings to mind the phrase, "Rome burned while Nero fiddled," except in this case it is not just a city, or even a country, that is at stake. It is literally the survival of the vast and integrated global ecosystems that allow humans to survive in an otherwise hostile universe.

Two of the key conclusions of this book are that humanity is as close to a technological and biological utopia as it is to oblivion, and it is possible to have prosperity without pollution -- if responsible changes are implemented in time.

Given the power of the electronic and print media to focus the world's attention on key events, there is every reason to be hopeful that the necessary changes can be made in time. Consider the media attention that was given to President Clinton's sexual exploits, the Exxon oil tanker spill in Alaska, or the cold fusion reaction by chemists at the University of Utah. The cold fusion event was particularly interesting because most news reporters and Members of Congress were apparently surprised to learn that someone may have discovered a process that could provide unlimited energy without producing significant amounts of radioactive waste, greenhouse gases, or acid rain. What is hard to understand is that the renewable hydrogen energy systems reviewed in this book have the same advantages of the cold fusion nuclear process, but the print and electronic media have yet to "discover" them.

The hydrogen technologies are not unproven theoretical curiosities. Rather, they are well-characterized technologies that are being commercialized by a number of major companies worldwide. Although many people assume that major oil companies would never allow renewable energy systems to be implemented, the reality is that many oil companies, such as Shell, British Petroleum (BP)/Amoco and Texaco have already established hydrogen business units. As one oil executive from BP explained, "the transition to a hydrogen economy is inevitable, the only question is one of timing." Unlike the traditional high-risk investments required to find the remaining oil reserves, renewable energy systems will provide a renewable rate of return with little risk, and without any of the problems associated with contaminating the Earth's biological life-support systems.

Because the transition to a hydrogen energy system will solve many of the most serious energy and environmental problems, it is a transition that both environmentalists and oil company executives will be able to support. This is an important consideration, because few other significant issues could attract the support of such divergent interest groups. This being the case, many people ask an obvious question: If the solar-hydrogen energy system has so many advantages, why don't the oil companies just start making hydrogen?

The question is understandable. The answer is two-fold. First, a substantial information-gap exists even between energy engineers. Because of the information explosion, scientists and engineers must increasingly specialize, and as a result, they can only be knowledgeable in relatively narrow fields. Even investigators in the various solar and wind energy areas typically only have a limited awareness of the hydrogen engineering community; and likewise, many of the hydrogen engineering specialists are unaware of the most significant renewable energy technologies that could be mass-produced for large-scale hydrogen production. If most engineering analysts have not put all of the pieces together, it is not surprising that most elected officials have not seen "the big picture."

In addition to this information gap, which is compounded by the exponential knowledge explosion, there is also the fact that individual companies do not determine national energy policy. Even if all of the oil companies were ready to make the transition to hydrogen, they would still need to coordinate this transition with the automobile and aircraft companies that would all have to follow an agreed upon timetable in terms of modifying their vehicles to use hydrogen fuel. Other infrastructure systems, including pipelines, ships and hundreds of other manufacturers that make everything from stoves to water heaters must also be involved in the transition timetable. As such, an energy transition from oil to anything else is an undertaking that cannot be implemented without government leadership, coordination and support. In the same way that industry cooperated with the War Production Board in World War II, the President now needs to set up a Hydrogen Production Board to coordinate a national and international industrial transition from oil to hydrogen.

Political Priorities

In spite of the many environmental and scientific warning signs, there has been an abysmal lack of political leadership. Members of Congress have demonstrated a "business-as-usual" attitude that clearly indicates they or their senior advisors are not aware of the issues related to shifting from oil to hydrogen. Indeed, there is the continued insistence to promote nuclear power and to extract the last of the oil offshore and in the remaining wilderness areas, including the Artic National Wildlife Refuge in Alaska, where the reserves would only be able to sustain the U.S. for from 6 months to a few years at best.

Some Members of Congress have promoted the use of ethanol as an automotive fuel to replace gasoline, but according to John Appleby, director of Texas A&M's Center for Electrochemical Systems, ethanol, methanol and other alcohols derived from grain cannot contribute very much toward solving the air-pollution or climate change problems. Although such fuels have only half the carbon content of gasoline, Appleby points out that it takes twice as much fuel to travel the same distance, which means the carbon content emitted to the atmosphere is about the same as gasoline for the same energy content. Thus, making a transition to such "alternative fuels" accomplishes little.

Although the Clinton Administration was generally viewed as supportive of environmental issues and acknowledged that the global environmental problems are serious, its policies were woefully inadequate. It seems clear that major, rather than minor "band-aid" changes, are called for. In acknowledgement of the seriousness of the global environmental crisis, *Time* magazine altered its normal policy of selecting a person of the year in 1989, and instead highlighted the fact that the Earth is in serious trouble. The following is excerpted from one of the main articles titled, "What on Earth Are We Doing?"

"Let there be no illusions. Taking effective action to halt the massive injury to the Earth's environment will require a mobilization of political will, international cooperation and sacrifice unknown except in wartime...As man heads into the last decade of the 20th century, he finds himself at a crucial turning point: the actions of those now living will determine the future, and possibly the very survival of the species. Now more than ever, the world needs leaders who can inspire their fellow citizens with a fiery sense of mission, not a nationalistic or military campaign but a universal crusade to save the planet. Unless mankind embraces that cause totally, and without delay, it may have no alternative to the bang of nuclear holocaust or the whimper of slow extinction."

While many people understandably mistrust governmental regulatory agencies, the fact remains that these are the only institutions that can require private companies to comply with the regulations that are deemed to be in the public interest.

Government Regulation

Many members of the U.S. Congress are very much opposed to environmental regulations, and the governments of many other countries are further behind the U.S. with respect to having -- much less enforcing -- environmental guidelines. The sad reality is that at a time of the greatest peril in history, there seems to be no one at the helm. The problem seems to be that the President, the Members of Congress, the general public, and the media which generally informs them, are not aware of a specific plan that can resolve the multiple and interrelated energy and environmental problems. After all, being aware of the problems is one thing. Being aware of the solutions is quite another.

Even those individuals who take the time to be informed are likely to throw up their hands and say, *If the "experts" don't have a plan, how is the average citizen supposed to come up with one?* But as this book has documented, there are solutions to the energy and environmental problems. Indeed, humanity is as close to utopia as it is to oblivion, and whether one or the other evolves will ultimately depend on citizen action. Elected officials generally follow public opinion polls with great care, for their political survival depends upon serving the interests of their constituents. What this means, however, is that elected officials are by definition followers and not leaders. The leaders are, therefore, the majority of voters, and until they demand that their governmental representatives make a transition from non-renewable fossil fuels, such as oil, to renewable hydrogen, it is highly unlikely that such a fundamental change will ever happen.

Although there are a great many people who sincerely believe that government regulation is the problem and a less regulated economy is the solution; it is well to remember that many of the most serious environmental problems have been the result of a lack of governmental regulation. Indeed, most regulations are only enacted after serious problems have already occurred. History has repeatedly shown that a true free enterprise system requires a reasonable amount of government regulation if it is to remain free. There is no better example than what happened with John D. Rockefeller or more recently, Bill Gates. When these hard working entrepreneurs started out, their respective areas of business were highly competitive. But over time, both Rockefeller and Gates became more and more successful than their competitors, and they were eventually able to buy them out, and in so doing, created a monopoly that effectively ended the competitive free market forces. Thus, an unregulated free market contains the seeds of its own destruction.

While governmental agencies are far from perfect, they are the only institutions that are responsible for looking after the health and welfare of the general public. As a result, a more balanced and realistic approach would involve establishing a partnership between the private sector and the governmental regulatory agencies. Using the analogy of a football game, private companies should be like the players in the game while the government officials are the referees, who establish and enforce the rules. As long as everyone competes fairly under the established rules, there is no reason for the governmental agencies to interfere with the corporate players.

The importance of governmental officials is that they are supposed to be somewhat objective and represent all of the people. This is in contrast to individuals working for private companies who are expected to look out for their company's self-interest. It is not an "either-or" question of whether an unregulated economy is inherently bad or good. It is a matter of acknowledging how a profit-oriented free-market system works. Private companies cannot be expected to take on social responsibilities that are not in their own short-term economic self-interest. While most companies seek to cooperate with the government, few -- if any -- want to assume the responsibilities of the government to protect the public welfare. It is simply not their function.

Acknowledging the inherent nature of companies to focus on short-term profitability at the expense of long-term planning, as well as research and development, Japan has developed a highly successful Ministry of International Trade & Industry (MITI) that has been responsible for coordinating Japan's long-term industrial strategy. In sharp contrast, U.S. companies essentially make business decisions in isolation. Corporate executives rarely have input from governmental officials on important corporate business decisions, and most companies would never think of consulting with their competitors on such matters, which in many cases, may actually be illegal in the U.S. Japan, on the other hand, has been able to evolve into an economic and technological superpower by having its private companies work in partnership with MITI.

One thing is clear. The magnitude and severity of the problems confronting the human community underscores the need for a balanced approach to optimizing industrial and environmental policy decisions, where cooperation is valued as well as competition. When the industrial transition to hydrogen technologies and resources gets the kind of intense media exposure that the alleged cold fusion experiments received, the American public will demand that the transition begin as soon as possible. It is hoped that this book will help to advance this critical energy and industrial transition before the global environmental and economic disruptions make such a transition impossible. Given the exponential nature of the global energy and environmental problems, it is time for elected officials to shift from a policy of attempting to *manage* the problems, and to take the fundamental steps necessary to *solve* them.

On a federal level, the U.S. government needs to establish a national energy policy that will allow the industrial transition to renewable resources to occur as rapidly as possible. While some people believe it will take 50 or 100 years for the U.S. to make a transition from oil to a hydrogen energy system, it is well to recall that in World War II, every major industry in the U.S. was retooled in 12 months. In the same way that the War Production Board was established in World War II to coordinate the industrial retooling effort, a similar Hydrogen Production Board needs to be established to coordinate the energy, automotive and aerospace companies that will all be directly involved in the effort. Moreover, the net result of this effort will be good for the overall economy because millions of manufacturing jobs will be created in the process. While a Hydrogen Production Board is not presently being considered by anyone in the U.S. Congress, more moderate bipartisan hydrogen development legislation has been passed in both the House and the Senate.

Fair Accounting Act

One of the more significant first steps that needs to be taken on both a federal and state level in the U.S. is for a fair accounting system to be established that will factor in the external costs of using fossil fuels. The language for a Fair Accounting Act could be worded as follows:

PROPOSED

UNITED STATES LEGISLATION

Fair Accounting Act

Whereas it is well documented that the vast majority of the increasingly unhealthy urban air pollution in major metropolitan areas is a direct result of the combustion of carbon-based fuels for automobiles, trucks, aircraft and power plants, and;

Whereas carbon-based fuels are non-renewable and therefore unsustainable fuel options that must be imported into the U.S., and;

Whereas hydrogen is the only zero carbon emission combustion fuel that can be made from both fossil fuel and renewable resources, and;

Whereas all green plants on the earth have been utilizing a solar-hydrogen energy production process for over 3 billion years whereby the hydrogen is extracted from water with solar energy, and;

Whereas hydrogen can allow the United States to become energy independent and essentially pollution-free, and;

Whereas the United States has unique land and solar energy resources that would mean the United States has the potential to develop into a "Saudi Arabia"-class energy exporter of a pollution-free hydrogen fuel that is inexhaustible, and;

Whereas 18 gallons of water is required to manufacture a gallon of gasoline, and only 2.3 gallons is required to make a similar energy content of hydrogen, and;

Whereas NASA, Boeing, Lockheed and Airbus have documented that hydrogen can be safely used to fuel commercial aircraft and spacecraft, and;

Whereas major oil companies, such as Texaco, Shell and BP/Amoco have already established hydrogen business units, and;

Whereas hydrogen is safer in the event of accidents than hydrocarbon fuels such as gasoline, or natural gas, and;

Whereas investigations published in Scientific American have shown that the "external costs" of carbon based fuels, which include health care costs, corrosion to buildings and bridges, military costs, crop losses, employment and subsidies cost U.S. citizens up to $300 billion annually; and;

Whereas this $300 billion in external costs does not factor in the fact that the vast majority of U.S. citizens are forced to live and raise their children in highly polluted areas that seriously degrade the quality of life on a day-to-day basis, and;

Whereas the United States consumes about 200 billion gallons of gasoline and diesel fuel annually;

Be It Resolved, therefore, that a Fair Accounting Act shall provide that $1.00 per gallon be gradually assessed, starting at $0.10 per gallon on January 1, 2001, and increasing each year by an additional $0.10 until January 1, 2010. Non-renewable gaseous fuels containing carbon, such as natural gas or propane, shall be assessed on a comparable Btu basis as liquid hydrocarbon fuels.

Be It Further Resolved that the funds raised by the Fair Accounting Act shall be used to provide low interest long-term financing for the capital-intensive carbon-free solar hydrogen energy systems that will be needed to displace the non-renewable carbon fuel sources that must now be imported and used within the United States. It is anticipated that individual State Legislatures will also impose similar financial incentives on a state level to provide a level playing field for the renewable energy technologies.

Given that the renewable energy technologies typically do not have fuel or waste storage costs, a major component of their electricity costs are the "up front" capital cost requirements for building the systems. As a result, the renewable energy technologies are especially sensitive to the interest costs that are used to finance their construction. This being the case, the financial resources collected by the Fair Accounting Act could be used to provide low cost financing (in the range of one or two percent interest), which will significantly reduce the cost of the electricity and/or hydrogen that will be produced. While some people will oppose such special consideration being provided to renewable energy technologies, it is well to recall that over a trillion dollars has been spent by the United States in the development and maintenance of weapons of mass-destruction. It seems reasonable that providing a pollution-free and inexhaustible energy system deserves the same consideration. In the final analysis, virtually everyone will be impacted by such a transition.

It would seem that any reasonable person could understand the importance of protecting the Earth's biological life support system. Yet, the environmental habitats of the Earth are being destroyed at a terrifying rate. As an old saying goes, *"We have found the enemy, and it is us."* Two of the more important considerations are that it is critical to take action while there is still time to make a difference, and that major -- and not minor -- changes are called for. Moreover, if the leadership in changing national priorities is not provided by elected officials within the state and national governments, then it must come from the public at large. However, before such a public initiative is going to occur, the public needs to be made aware that there are viable solutions to the most serious problems that we collectively face, and that the essence of the solutions involves making an industrial transition to the types of renewable energy resources and technologies that have been documented in this book.

What is needed is a Moonshot or Manhattan Project type program to shift to the hydrogen energy system, led by an integrated team of committed professionals who understand that that this transition needs to be implemented with wartime speed.

Fortunately, much work has already been accomplished in terms of identifying the renewable energy technologies that can now be mass-produced for large-scale hydrogen production, and major automotive companies, including BMW, DamilerChrysler,

Mazda, Ford and General Motors are all developing hydrogen-fueled cars. However, the contributions made by investigators at BMW are particularly noteworthy. Although there are virtually no hydrogen refueling stations available, BMW has been committed to optimizing vehicles to operate on both liquid hydrogen and gasoline fuel for more than 20 years. Their current fleet of these pre-production "bifuel" or "bivalent" cars, as the BMW hydrogen engineers refer to them, are also integrated with a fuel cell in the trunk that serves as an on-board hydrogen-fueled electrical generator. The only difference that is noticeable about the BMW hydrogen-fueled cars is that they have two fuel caps; one for liquid hydrogen and one for gasoline.

According to Wolfgang Strobl and Hans-Christian Fickel, two of the senior engineers at BMW that have been involved in the hydrogen car project, there are no obstacles to the mass-production of the hydrogen-fueled cars. All of the key components are made of common materials, such as steel and aluminum. Both Fickel and Strobl were on hand at the Clean Energy Exhibition that was held in Munich, Germany from June through September of 2000. This multi-million dollar event was in part sponsored by BMW, and their President and CEO, Professor Joachim Milberg, was the main speaker at the grand opening. His opening remarks included the following:

"We have invited you here so that together we can begin a new era, an era of completely emission-free vehicle power. In short, you are witnesses to the fact that we have succeeded in bringing a long-time research goal to reality. Therefore, I am proud to say that the BMW Group has done significant pioneering work in the development of hydrogen power.

Today, we are able to say to you, "Get in -- take a seat in our hydrogen car. You can 'experience' the future with us -- in the truest sense of the word. What we are showing you in this exhibit is not a study, . . .What we are showing you today is already in small-scale production. If you look underneath the hood of our hydrogen vehicles, you will find that, outwardly, they are not significantly different from those equipped with gasoline-fueled engines. This is exactly what is fascinating about them. They are completely 'normal' cars.

Mobility is not a desire created by automobile manufacturers to promote the sale of our automobiles. . . mobility is, plain and simple, a basic need that people have.

One thing is clear, and on this matter there is a consensus across all party lines: our fossil fuels are limited. Moreover, when they are burned, carbon dioxide is created, which is partly to blame for the compounding of the greenhouse effect. Thus, what we need are alternatives that save energy and protect the climate; and the sooner the better.

Our answer is cars powered by Sun and water.

The fuel of the future will be hydrogen. Hydrogen is the absolute cleanest alternative to fossil fuels and hydrogen in the form of water can be found on Earth in almost inexhaustible quantities.

As you know, some automobile manufacturers favor fuel cell batteries for electric vehicles. We say this: we will use fuel cells in the vehicle only where electricity is used directly -- mainly due to the high cost and weight of fuel cells. As you know electrical demand continues to increase for the many auxiliary functions in the vehicle, and a hydrogen-powered fuel cell that is incorporated in the hydrogen-fueled BMW can power stationary heating and/or air conditioning when the engine is not running. This allows the fuel cell battery to assume only that function which it best fulfills; namely the efficient production of electricity.

One thing is getting more and more clear; alternatives such as the electric car, involve so many limitations that the customer is not accepting them. Ten years ago the electric car was still being promoted as the answer, but what followed was too little power, a lot of additional weight, and an unusually small range.

If we want to achieve acceptance, we need a drive technology that takes our customers' needs seriously. We don't want 'sacrifices on wheels' as one journalist so graphically put it. Therefore, our strategy is to rely on hydrogen power, which provides sufficient power and range: and all of this -- I can only repeat this -- completely emission-free."

Professor Joachim Milberg
President & CEO
BMW Group

Figure 9.4: BMW Hydrogen Fleet
The manufacture of the BMW "hydrogen 7 series" cars included a five
kilowatt fuel cell, also fueled by hydrogen, that generates electricity for
heating and air conditioning even when the engine is not running.

Figure 9.5: Clean Energy Exhibition
An indoor view of the Clean Energy Exhibition in Munich shows a liquid hydrogen-fueled BMW with a transparent hood that reveals the V-
12 engine that can operate on either hydrogen or gasoline.

Figure 9.6: BMW 750 hL
The 5th generation BMW 750hL is the first series production hydrogen car in the world, and BMW is now taking orders for such cars.

Figure 9.7: Hydrogen Pioneers
Wolfgang Strobl (left) who is referred to as the father of the hydrogen car program for BMW and Hans-Christian Fickel (center) who directed the hydrogen-fueled engine development program for BMW are two of the senior level engineers that have been highly involved in the hydrogen car project. They were both in attendance with the author (right) for the grand opening of the Clean Energy Exhibition in Munich, Germany.

In the case of BMW, the high-level of enthusiasm for the hydrogen-fueled car program and the more general aspects of the transition to a hydrogen energy system, seems to be a part of the corporate culture. BMW has now built the world's first fleet of series production liquid hydrogen-fueled V-12 powered automobiles, which have a range with the liquid hydrogen tank of roughly 250 miles). Referred to as the 750hL, the only significant modifications to the engine concern the fuel intake ports, which have additional injector valves for hydrogen as well as for gasoline. An initial fleet of 15 of the "hydrogen 7 series" BMWs is being operated on a daily basis in both Hannover and Munich.

BMW has also participated as part of a corporate team in Europe which has developed the world's first commercial gaseous and liquid hydrogen fueling station that is located near the main terminal of the Munich International Airport. The station is equipped with its own electrolyzer system to manufacture the gaseous hydrogen from water and electricity, and the corporate team that developed the system also includes Linde, Siemens, MAN and Aral. Both automobiles as well as the airport buses are refueled at the hydrogen pumping station, which is also equipped with a robot that automatically finds the liquid hydrogen tank fuel cap and then refuels the BMW vehicles with liquid hydrogen in about 4 minutes. The driver merely swipes his credit card and waits in the vehicle until the process is complete. What BMW has clearly demonstrated is that hydrogen-fueled vehicles are indeed practical for mass-production and use. It is also significant to see both Shell Oil Company and British Petroleum (BP)/Amoco supporting the general notion that the transition to a hydrogen energy system is not a question of whether -- but when.

As Chapter 6 indicated, the major infrastructure issue that needs to be resolved has to do with the need for a greatly expanded electricity production capability. The U.S. currently consumes about 95 quadrillion Btu of energy annually, and in order to provide this much energy with electrolytic hydrogen production systems, roughly 10 million megawatts of electrical generation would need to manufactured and installed. Given the scale and scope of such an effort, only the renewable energy technologies have the potential to be mass-produced for such a task. Some of the important questions are: *What technologies will be the most cost-effective for large-scale hydrogen production? How much will this project cost and how long will it take?*

The Phoenix Project

Assuming wind systems (both onshore and offshore) were to be used as a case in point, roughly 10 million one-MW wind turbines would need to be built to make the United States energy independent. While that may seem like a lot of wind turbines, consider that from a material and manufacturing perspective the electrical power conversion unit of a wind turbine is in similar to a large automobile or truck, and over 15 million automobiles and trucks are manufactured in the U.S. each year! Current wind energy systems that are in relatively small production runs have an installed capital cost of approximately $1,000 per installed kilowatt (which results in an electricity cost of about 5 cents per kWh). If the wind machines were mass-produced on a scale to displace fossil fuels, it would be reasonable to assume that these capital and electricity costs would be reduced by as much as 50 percent, resulting in electricity costs would be expected to be reduced accordingly, to approximately 2-cents per kWh. Moreover, if long-term low interest financing were provided as a result of the Fair Accounting Act, or comparable investment tax credits are provided by the U.S. Congress, the electricity costs could be reduced to less than 1-cent (10 mills) per kWh.

According to a 1991 study prepared for the Bonneville Power Administration by Fluor Daniel Corporation, with electricity costs of 1-cent per kWh, gaseous hydrogen could be commercially produced for about $15.00 per million Btu. If the hydrogen is to be liquefied, an additional $4.00 per million Btu would be added, bringing the total cost to about $19.00 per million Btus. This is equivalent to gasoline at the refinery costing about $2.30 per gallon. Assuming taxes and distribution costs are also factored into the cost of liquid hydrogen, it would be reasonable to expect that it would cost about $3.00 per equivalent gallon of gasoline at the pump. This underscores the point that if a fair accounting system were in place, hydrogen would be the least expensive fuel. There is also the consideration that the cost of hydrogen produced by renewable energy technologies will always be coming down in the future, compared to gasoline and other hydrocarbon fuels that will only be getting more expensive as the global fossil fuels reserves are depleted.

In the June 24th, 2000 edition of *The Arizona Republic,* journalist Steve Wilson made the following observations:

"Current polls show many voters regard high gasoline prices as the country's Number 1 economic problem and both (Presidential) candidates have been eager to exploit the issue. You do not hear either man leveling with us and saying today's higher prices follow two decades of ignoring our country's increasing reliance on foreign oil.

In 1973, when the U.S. imported 35 percent of its oil, Richard Nixon set a goal of energy self-sufficiency by 1980, and Congress provided subsidies for developing alternative fuels. The subsidies didn't last, and U.S. public and private investment in energy research has dropped by more than half. As a result, 55 percent of our oil now comes from abroad.

If Jimmy Carter asked too much when he tried to rally the nation around his energy plan, today's leaders ask too little. As long as American's insist on consuming more without investing to produce more, we shouldn't be shocked to see prices climb."

Figure 9.8: U.S. Energy Policy

Figure 9.9: Gas Lines
While gasoline is something we take for granted, without it, the lifestyle
begins to change profoundly -- and usually for the worse.

Figure 9.10: Instability
Civilization is paper thin, and when it tears, it tears in a matter of hours. If either energy or food production systems fail, history has shown that chaos can follow very quickly.

Wilson points out in his article that in countries such as Germany, Japan and Britain, gasoline prices have been between $3.50 to $4.80 for years. Although the higher gasoline fuel prices in these countries are due to taxes, it would be hard to argue that the quality of life in these countries is any less that is typically found in the U.S. However, according to Phil Verleger, an energy analyst for the Boston-based Brattle Group, assuming high gasoline prices do not create a worldwide economic recession, demand for oil will soon increase by another two million barrels per day, and such an increase can only be supplied by Saddam Hussein's Iraq, where exports have been curtailed because of international sanctions. As Verleger characterized it:

"At that point, I think Saddam Hussein will have the world where he wants it."

This chilling scenario underscores the important interrelationships that exist between energy and the economy, as well as international political relations.

The more important question is: *What will the price of gasoline will be in 2005 or 2010, given that the world is consuming four barrels of oil for every one that is produced?* Given these observations, it is clear that it is possible to have the U.S. converted to a renewable hydrogen energy system by the year 2010 -- if not sooner. It is only a matter of public awareness that must be translated into governmental policy. Consumers now have the power to significantly accelerate the transition to the hydrogen energy system by placing an order a hydrogen-fueled car that also has the capability to operate on gasoline until such time that hydrogen pumps are in widespread use.

If enough people sign up for bivalent vehicles, the market will respond. BMW is ready to deliver such vehicles with all the quality and reliability that one expects to have with a conventional gasoline-fueled car. These dual fueled, "bivalent" cars are slightly more expensive than mono-fueled vehicles, but the marginal cost increases can be significantly reduced in mass-production. In small production runs, the additional fuel tank and storage system increases the price of the car by about 25 percent. But if such cars were manufactured in high volumes, the cost increases would only raise the vehicle cost by about 5 percent. Such minimal cost increases are offset by the impact such vehicles would have in accelerating the transition to clean energy technologies. It is not necessary to have another generation of children grow up in polluted cities. If consumers place their orders now, they will help insure that we will not have to wait until 2050 to have prosperity without pollution.

It is hoped that this book will help to advance this critical energy and industrial transition before the global environmental and economic disruptions make such a transition much more expensive -- if not impossible. Given that a large number of contaminants have already been released into the global environmental systems, it is now likely that a number of ecological system collapses will occur. This will significantly reduce the existing human population levels, and most of the Earth's remaining ancient wilderness areas. But human adaptability being what it is, it is reasonable to expect that with a successful energy transition to renewable resources and the rapid expansion of controlled-environment agricultural systems, human civilization will indeed be able to continue. This, in turn, will ensure that the computer and biotechnology advances will continue as well.

Given that medical researchers in biochemistry and molecular biology have already begun to regenerate organs and other biological tissue in mammals, it is now only a question of time before humanity makes a biological transition to renewable resources. With the "designer gene" era described in Chapter 8 rapidly, indeed -- exponentially -- approaching, whereby disease and aging will be virtually eliminated, it would be ironic indeed if the civilization that allowed such truly astonishing developments to occur disintegrated because it also allowed its own biological life-support systems to be destroyed.

The latest reports from NASA regarding stratospheric ozone-depletion are not encouraging. As of April 2000, the Earth's protective ozone layer has thinned to record low levels around the Arctic region. This is in spite of the international treaty that was enacted in 1996 to ban ozone depleting chemicals. Moreover, according to research that was presented to a recent American Geophysical Union meeting in Washington D.C., the thinning ozone layer may be headed for even more dramatic losses because of global warming issues. Apparently, the increasing surface temperatures in the lower atmosphere leads to a cooling effect in the upper atmosphere.

A newer study, published in the May 2000 issue of the journal *Science*, examined the role of polar clouds that form in the stratosphere, which is approximately 9 to 25 miles above the surface of the Earth. Unfortunately, these polar clouds provide a surface where a chemical reaction changes harmless forms of chlorine into their ozone-destroying cousins. These troublesome clouds are persistent features of the Antarctic stratosphere, where temperatures are typically colder than they are in the Arctic region. According to Azadeh Tabazadeh, an atmospheric chemist at the NASA Ames Research Center in Mountain View, California who led the study, *"the Arctic could soon become 'Antarctic like' -- colder and filled with polar clouds for longer periods."* Michele Santee, an atmospheric scientist at NASA's Jet Propulsion Laboratory in Pasadena, California who co-authored the study, made the following observations:

> *"We do know that if the temperatures in the stratosphere are lower, more polar clouds will form and persist, and these conditions will lead to more ozone loss."*

In addition, the large polar clouds are formed from condensed nitric acid and water. When they become large, the nitrogen that normally helps to stop ozone loss falls out of the clouds as snow. Scientists say their new research makes them fear that the ozone layer is not able to recover as fast as they had once thought. In some parts of the upper atmosphere, up to 60 percent of the protective ozone layer has been lost.

In the exponential terms that were discussed in Chapter 2, that makes it 11:59; one minute before midnight.

Given these considerations, it could be clear that humanity now stands at the threshold of the end of life as we know it, and given the exponential nature of the events now unfolding, the oblivion and/or utopia scenario may occur sooner than most people expect. Utopia in this case is not defined as some general philosophy, but the atomically precise control and understanding of the physical and chemical world in which we live, which of course includes our own unique genetic and molecular structure. While not every one will choose this path, from the perspective of this observer, it is the point of the physical and chemical evolution of the universe that has been going on since the big bang occurred some 15 billion years ago. As Carl Sagan has written, human beings are the products of 15 billion years of hydrogen atoms evolving in the universe. As such, we and our biocybernetic decedents are the meaning and the purpose of life, for we have come to understand the very essence of life and matter in the known universe. That is reason enough for being.

There will always be some people who will argue that elimination of aging and other diseases -- much less the development of "designer genes" and the evolution of an immortalist culture of biocybernetic organisms -- is immoral or otherwise too risky. The reality is, however, that no one is really in charge of these technology developments. Technologies have a life of their own. There are thousands of investigators that are working long hours to advance the interdisciplinary fields of chemistry, molecular biology, computer science and nanotechnology. As such, assuming a shift from oil to hydrogen is accomplished and the miracle of civilization is allowed to continue, an immortalist culture whereby *Homo sapiens* evolves into *Homo immortalis* is inevitable.

The stakes could not be higher.

This reality underscores the sense of urgency to demand decisive action by our elected representatives.

If there are those who say they have a better solution to the global energy and environmental problems than making an industrial transition from the oil economy to the hydrogen economy, let them come forth and demonstrate a viable alternative. If not, then let's get on with the job that needs to be done. We have waited long enough, and the opportunity to stand and be counted is rapidly slipping away.

*   *   *

REFERENCES

CHAPTER 1:
UTOPIA OR OBLIVION

1.   J. O'M. Bockris, "Hydrogen Economy," SCIENCE (American Association for the Advancement of Science), Vol. 176, No. 4041, p. 1323, June 23, 1972.

2.   Derek P. Gregory (Institute of Gas Technology), "The Hydrogen Economy," SCIENTIFIC AMERICAN, Vol. 228, No. 1, pp. 13-21, January 1973.

3.   T. Nejat Veziroglu and A. N. Protsenko, Hydrogen Energy Progress VII: Reviewing the Progress in Hydrogen Energy, PERGAMON PRESS (New York, New York), October 1988.

4.   J. Pangborn, M. Scott and J. Sharer, "Technical Prospects for Commercial and Residential Distribution and Utilization of Hydrogen," INTERNATIONAL JOURNAL OF HYDROGEN ENERGY (International Association for Hydrogen Energy), Vol. 2, pp. 431-445, 1977.

5.   Tom Wicker, "Chilling Thoughts of Summer," THE NEW YORK TIMES release, published in the ARIZONA REPUBLIC (Phoenix, Arizona), p. A13, January 17, 1989.

6.   Mary Wayne, et al., "Acid Rain: Clarifying the Scientific Unknowns," ELECTRIC POWER RESEARCH INSTITUTE (EPRI) JOURNAL, Vol. 8, No. 9, p. 8, November 1983.

7.   Volker A. Mohnen, "The Challenge of Acid Rain," SCIENTIFIC AMER-ICAN, Vol. 259, No. 2, pp. 30-38, August 1988.

8.   "Caustic fog poses worse threat to health, ecology than acid rain," LOS ANGLES TIMES release, published in THE ARIZONA REPUB-LIC, p. A5, November 12, 1982.

9.   "Acid-fog hazard is feared worse than announced," UNITED PRESS INTERNATIONAL release, published in THE ARIZONA REPUBLIC, p. A5, November 15, 1982.

10.   Robert W. Shaw (chief of chemical diagnostics & surface science at the U.S. Army research office in Research Triangle Park, North Carolina), "Air Pollution by Particle," SCIENTIFIC AMERICAN, Vol. 257, No. 2, pp. 96-103, August 1987.

11.   Richard A. Houghton and George M. Woodwell (Woods Hole Research Center, Woods Hole, Massachusetts), "Global Climatic Change," SCIENTIFIC AMERICAN, Vol. 260, No. 4, p. 36, April 1989.

12.   Philip Shabecoff, "Greenhouse effect here, expert fears," NEW YORK TIMES release, published in THE ARIZONA REPUBLIC, p. A1, June 24, 1988.

13.   Syukuro Manabe and Richard T. Wetherald (National Oceanic and Atmospheric Adminstration's Geophysical Fluid Dynamics Laboratory, Princeton, University)

14.   "Reduction in Summer Soil Wetness Induced by an Increase in Atmospheric Carbon Dioxide," SCIENCE, Vol. 232, No. 4750, pp. 626-628, May 2, 1986.

15.   "Tampering with the Global Thermostat," COMPRESSED AIR MAGA-ZINE, Vol. 91, No. 10, pp. 16-22, October 1986.

16.   Irwin W. Sherman and Vilia G. Sherman, Biology: A Human Approach (Third Edition), OXFORD UNIVERSITY PRESS (New York, New York), pp. 599-600, 1983.

17.   "Tampering with the Global Thermostat," op. cit., p. 21.

18.   James E. Lovelock, Gaia: A New Look at Life on Earth, OXFORD UNIVERSITY PRESS (New York, New York), 1979.

19.   James E. Lovelock, The Ages of Gaia: A Biography of Our Living Earth, W.W. NORTON & COMPANY, INC. (New York, New York), 1988.

20.   James F. Kasting, Owen B. Toon and James B. Pollack, "How Climate Evolved on the Terrestrial Planets," SCIENTIFIC AMERICAN, Vol. 258, No. 2, pp. 90-97, February 1988.

21.   Gary Taubes, "Made in the Shade? No Way," DISCOVER, Vol. 8, No. 8, p. 65, August 1987.

22.   "Diminishing Ozone called Major Threat," THE ARIZONA REPUBLIC, p. A1, March 10, 1987.

23. "The Heat Is On," TIME MAGAZINE, Vol. 130, No. 16, p. 62, October 19, 1987.

24. Peter Aleshire, "Despite steps to halt it, pollution will still reduce ozone," THE ARIZONA REPUBLIC, p. AA2, April 10, 1988.

25. Karen E. Bettacchi, et al., "How Man Pollutes His World," NATIONAL GEOGRAPHIC MAGAZINE, 1970.

26. Linda Harrar, "The Hole in the Sky," NOVA, a Public Broadcast System (PBS) Science Documentary produced by the WGBH Educational Foundation, (Box 322, Boston, MA), 1987.

27. "Diminishing ozone called major threat: Skin-tumor death rate forecast to skyrocket," THE ARIZONA REPUBLIC, p. A1, March 10, 1987.

28. Richard J. Wurtman (Massachusetts Institute of Technology), "The Effects of Light on the Human Body," SCIENTIFIC AMERICAN, Vol. 233, No. 1, pp. 68-77, July 1975.

29. Robert M. Neer, T. R. A. Davis, A. Walcott, et al., "Stimulation by Artificial Lighting of Calcium Absorption in Elderly Human Subjects," NATURE, Vol. 229, No. 5282, pp. 225-226, January 22, 1971.

30. A.J. Lewy, et al., "Light Suppresses Melatonin in Humans," SCIENCE, Vol. 210, No. 4475, pp. 1267-1269, December 1980.

31. "Ozone 'worse than acid rain' in short term," BOSTON GLOBE release, published in THE ARIZONA REPUBLIC, p. A4, December 26, 1985.

32. "Nation's water tainted by leaks in gasoline tanks," NEW YORK TIMES release, published in THE ARIZONA REPUBLIC, p. A1, November 30, 1983.

33. "One million Americans likely to get cancer in '89, report says," UNITED PRESS INTERNATIONAL release, published in THE ARIZONA REPUBLIC, p. A10, February 25, 1989.

34. Paul Raeburn, "Aid, not drought, caused African Famine," ASSOCIATED PRESS release, published in THE ARIZONA REPUBLIC, p. A3, November 20, 1985.

35. Jeff Nesmith, "Evidence shows Earth heating up," COX NEWS SERVICE, published in the MESA TRIBUNE (Mesa, Arizona), p. A20, November 27, 1986.

36. William D. Ruckelshaus, "Toward A Sustainable World," SCIENTIFIC AMERICAN, Vol. 261, No. 3, pp. 166-174, September 1989.

CHAPTER 2:
EXPONENTIAL ICEBERGS

1.   Irwin W. Sherman and Vilia G. Sherman (University of California at Riverside), Biology: A Human Approach (Third Edition), OXFORD UNIVERSITY PRESS, p. 614, 1983.

2.   Albert A. Bartlett, "Forgotten Fundamentals of the Energy Crisis," AMERICAN JOURNAL OF PHYSICS (American Association of Physics Teachers), Vol. 46, No. 9, pp. 876-888, September 1978.

3.   Sherman, et al., p. 614.

4.   Carl Sagan, Cosmos, RANDOM HOUSE (New York, New York), p. 281, May 1980.

5.   Alan Toffler, Future Shock, RANDOM HOUSE (New York, New York), p. 26, 1970.

CHAPTER 3:
FOSSIL FUELS

1.   Annual Energy Review 1998, published by the U.S. ENERGY INFOR-MATION ADMINISTRATION, for the U.S. Department of Energy, Washington, D.C.

2.   Dr. M. King Hubbert, "The Energy Resources of the Earth," SCIEN-TIFIC AMERICAN, Vol. 224, No. 3, pp. 66-70, September 1971.

3.   M. K. Hubbert, "Energy Resources: A Report to the Committee on Natural Resources," NATIONAL ACADEMY OF SCIENCES--NATION-AL RESEARCH COUNCIL, p. 1, Publication 1000-D, Washington D.C., 1962.

4.   Harrison Brown, "Energy In Our Future," reprinted by John C. Bailor, Jr., et al,. Chemistry, ACADEMIC PRESS, A Subsidiary of Harcourt Brace Jovanovich (New York, New York), p. 843, 1978.

5.   Edward H. Thorndike, Energy and Environment: A Primer for Scientists and Engineers, ADDISON-WESLEY PUBLISHING COMPA-NY (Reading, Massachusetts), p. ix, August 1978.

6.   M. A. K. Lodhi and R. W. Mires, "How Safe is the Storage of Liquid Hydrogen," INTERNATIONAL JOURNAL OF HYDROGEN ENERGY, Vol. 14, No. 1, p. 36, 1989.

7.   Donald F. Othmer, "Energy -- Fluid Fuels from Solids," MECHANICAL ENGINEERING (American Society of Mechnical Engineers), Vol. 99, No. 11, pp. 29-35, November 1977.

8. Albert A. Bartlett (Department of Physics, University of Colorado), "Forgotten Fundamentals of the Energy Crisis," AMERICAN JOURNAL OF PHYSICS, Vol. 46, No. 9, p. 881, September 1978.

9. Charles D. Masters, "World Petroleum Resources," Open-File Report No. 85-248, U.S. DEPARTMENT OF THE INTERIOR GEOLOGICAL SURVEY (USGS), September 1985.

10. Andrew R. Flower, "World Oil Production," SCIENTIFIC AMERICAN, Vol. 238, No. 3, pp. 42-49, March 1978.

11. "Fuel discovery outlook in U.S. dim, study says," NEW YORK TIMES release, reprinted in THE ARIZONA REPUBLIC, p. A1 & A17, Sunday, April 12, 1981.

12. Allen E. Murray (Mobil Oil Corporation), "The Impending Energy Crisis," NEWSWEEK MAGAZINE, Vol. 105, No. 23, p. 16, June 10, 1985.

CHAPTER 4:
NUCLEAR POWER

1. Annual Energy Review 1998, published by the U.S. ENERGY INFORMATION ADMINISTRATION, for the U.S. Department of Energy, Washington, D.C.

2. S. Harwood, et al., "The Cost of Turning It Off," ENVIRONMENT, Vol. 18, No. 10, p. 17, December 1978.

3. Richard Severo, "Too hot to handle,"(cover story) NEW YORK TIMES MAGAZINE, p. 15, April 10, 1977.

4. Mark Trahant, "Idaho is Winner in Nuclear 'War,' " THE ARIZONA REPUBLIC, p. A8, February 11, 1989.

5. Mark Trahant, "Storage Plan Capsulizes Nuclear Worries," THE ARIZONA REPUBLIC, pp. A1-A8, February 11, 1989.

6. "Cancer Leaps Blamed on Chernobyl Fallout," ASSOCIATED PRESS release, reprinted in THE ARIZONA REPUBLIC, p. B7, February 19, 1989.

7. Richard Lipkin, "A Safer Breed of Reactor in Sight" (Integral Fast Reactor), INSIGHT, pp. 52-53, January 23, 1989.

8. John Horgan, "Fusion's Future," SCIENTIFIC AMERICAN, Vol. 260, No. 2, pp. 25-28, February 1989.

9.   "Fusion claims from Utah fail to gain backing," NEW YORK TIMES release, published in THE ARIZONA REPUBLIC, p. A11, April 29, 1989.

10.   Hearings, 95th Congress, "Nuclear Power Costs," HOUSE OF REP-RESENTATIVES, Report No. 95-1090, pp. 11-12, Committee on Government Operations, April 26, 1978.

11.   Barry Commoner, The Closing Circle, BANTAM BOOKS (New York, New York), p. 61, 1972.

12.   Kenneth and David Brower, "Miracle Earth," OMNI, Vol. 1, No. 1, pp. 16-18, October 1978.

13.   Draft Environmental Impact Statement for a Geologic Repository for the Disposal of Spend Nuclear Fuel a and High-Level Radioactive Waste at Yucca Mountain," published by THE U.S. DEPARTMENT OF ENERGY in July of 1999.

CHAPTER 5:
HYDROGEN

1.   Carl Sagan, Cosmos, RANDOM HOUSE (New York, New York), p. 233, 1980.

2.   Daniel L. Alkon, "Memory Storage and Neural Systems," SCIENTIFIC AMERICAN, Vol. 261, No. 1, pp. 44-45, July 1989.

3.   Carl R. Woese, "Archaebacteria," SCIENTIFIC AMERICAN, Vol. 244, No. 6, p. 100, June 1981.

4.   C. Marchetti, "From the Primeval Soup to World Government--An Essay on Comparative Evolution," INTERNATIONAL JOURNAL OF HYDROGEN ENERGY, Vol. 2, pp. 1-5, 1977.

5.   S. C. Huang, C. K. Secor, R. Ascione and R. M. Zweig, "Hydrogen Production by Non-Photosynthetic Bacteria," INTERNATIONAL JOURNAL OF HYDROGEN ENERGY, Vol. 10, No. 4, pp. 227-231, 1985.

6.   Michael D. Levitt, "Production and Excretion of Hydrogen Gas in Man," NEW ENGLAND JOURNAL OF MEDICINE (Massachusetts Medical Society), Vol. 281, No. 3, pp. 122-127, July 17, 1969.

7.   M. Bigard, P. Gaucher and C. Lassalle, "Fatal colonic explosion during colonoscopic polypectomy," GASTROENTEROLOGY (American Gastroenterological Association), Vol. 77, No. 6, pp. 1307-1310, December 1979.

8. Peter Hoffmann, The Forever Fuel: The Story of Hydrogen, WEST-VIEW PRESS (Boulder, Colorado), pp. 204-205, August 1981.

9. Irwin W. Sherman and Vilia G. Sherman, Biology: A Human Approach (Third Edition), OXFORD UNIVERSITY PRESS (New York, New York), pp. 599-600, 1983.

10. Abraham Lavi and Clarence Zener, "Plumbing the Ocean's Depths: A New Source of Power," IEEE SPECTRUM (Institute of Electrical & Electronics Engineers), Vol. 10, No. 10, pp. 22-27, October 1973.

11. J. O'M. Bockris, "Hydrogen Economy," SCIENCE, Vol. 176, No. 4041, p. 1323, June 23, 1972.

12. Derek P. Gregory (Institute of Gas Technology), "The Hydrogen Economy," SCIENTIFIC AMERICAN, Vol. 228, No. 1, pp. 13-21, January 1973.

13. Medard Gabel, Energy, Earth and Everyone, STRAIGHT ARROW BOOKS (San Francisco, California), p. 141, 1975.

14. "The Energy Crisis: One Solution," a film documentary produced by the College of Engineering, BRIGHAM YOUNG UNIVERSITY, 1976.

15. Walter Peschka (German Aerospace Research Establishment), "The Status of Handling and Storage Techniques for Liquid Hydrogen in Motor Vehicles," INTERNATIONAL JOURNAL OF HYDROGEN ENERGY, Pergamon Press, Vol. 12, No. 11, pp. 753-764, 1987.

16. G. Daniel Brewer (Lockheed Aircraft Corporation), "Hydrogen Usage in Air Transportation," INTERNATIONAL JOURNAL OF HYDROGEN ENERGY, Vol. 3, No. 2, pp. 217-229, 1978.

17. W. T. Mikolowsky and L.W. Noggle (The Rand Corporation), "The Potential of Liquid Hydrogen as a Military Aircraft Fuel," INTERNATIONAL JOURNAL OF HYDROGEN ENERGY, Vol. 3, No. 4, pp. 449-460, 1978.

18. Walter F. Stewart, "A Liquid Hydrogen-Fueled Buick," LOS ALAMOS NATIONAL LABORATORY (Los Alamos, New Mexico), Report No. LA-8605-MS, p. 7, November 1980.

19. "Liquid Hydrogen as a Vehicular Fuel," a report published by DEUTSCHE FORSCHUNGS-U. VERSUCHSANSTALT F. LUFT-U. RAUMFAHRT (i.e., the German Aerospace Research Establishment [DFVLR], Stuttgart, Germany 1982.

20. R. D. Quillian, et al., (U.S. Army Fuels & Lubricants Research Laboratory, San Antonio, Texas), Hydrogen Energy, Part B, PERGA-MON PRESS (New York, NY), p. 1025, March 1975.

21. Willis M. Hawkins and G. D. Brewer (Lockheed Aircraft Corporation), "Alternative Fuels Make Better Airplanes: Let's Demonstrate Now," ASTRONAUTICS & AERONAUTICS (American Institute of Aeronautics and Astronautics), Vol. 17, No. 9, pp. 42-46, September 1979.

22. Paul M. Ordin (NASA Lewis Research Center), "Review of Hydrogen Accidents and Incidents in NASA Operations," 9TH INTERSOCIETY ENERGY CONVERSION ENGINEERING CONFERENCE PRO- CEEDINGS, Technical Paper No. 749036, (American Society of Mechanical Engineers) pp. 442-453, August 1974.

23. B. C. Dunnam, "Air Force Experience in the Use of Liquid Hydrogen as an Aircraft Fuel," Proceedings of the FIRST WORLD HYDROGEN ENERGY CONFERENCE PROCEEDINGS, University of Miami (Miami, Florida), pp. 991-1010, March 1974.

24. A. M. Momenthy (The Boeing Commercial Airplane Company), "Fuel Subsystems for Liquid Hydrogen Aircraft: R & D Requirements," INTERNATIONAL JOURNAL OF HYDROGEN ENERGY, Vol. 2, pp.155-162, 1977.

25. Gerard K. O'Neill, "The colonization of space," PHYSICS TODAY, Vol. 27, No. 9, pp. 32-40, September 1974.

26. Gerard K. O'Neill, The High Frontier: Human Colonies in Space, BAN- TAM BOOKS (New York, New York), 1976.

27. Carl Sagan, Cosmos, RANDOM HOUSE, (New York, New York), p. 206, May 1980.

28. Howard P. Harrenstien (College of Engineering, University of Miami), "Hydrogen to Burn," OCEANUS (Woods Hole Oceanographic Institution),Vol. 17, pp. 28-29, Summer 1974.

29. Derek P. Gregory, et al. (Institute of Gas Technology), "The Economics of Hydrogen," CHEMTECH (American Chemical Society), Vol. 11, No. 7, pp. 432-440, July 1981.

30. C. R. Baker (Linde Division, Union Carbide Corporation), "Efficiency and Economics of Large Scale Hydrogen Liquefaction," SOCIETY OF AUTOMOTIVE ENGINEERS, Technical Paper No. 751094, Vol. 84, p. 132, November, 1975.

31. C. A. Rohrmann and J. Greenborg (Battelle-Pacific Northwest Laboratories), "Large Scale Hydrogen Production Utilizing Carbon in Renewable Resources," INTERNATIONAL JOURNAL OF HYDRO- GEN ENERGY, Vol. 2, pp. 31-40, 1977.

32. Willis M. Hawkins (Lockheed Aircraft Corporation), Letter to the Editor, INTERNATIONAL JOURNAL OF HYDROGEN ENERGY, Vol. 7, No. 1, p. 98, 1982.

33.   INTERNATIONAL ASSOCIATION FOR HYDROGEN ENERGY
        (IAHE), P.O. Box 248266, Coral Gables, Florida, 33124. T. Nejat
        Veziroglu, Editor in Chief.

34.   An editorial by Dr. T. Nejat Veziroglu (College of Engineering,
        University of Miami), "Hydrogen Energy System: Next Action,"
        INTERNATIONAL JOURNAL OF HYDROGEN ENERGY, Vol. 11, No.
        1, pp. 1-2, 1986.

35.   Carl Sagan, "Cosmos," PUBLIC BROADCASTING SYSTEM (PBS):
        Excerpted from the television documentary that was initially aired in
        1982.

CHAPTER 6:
RENEWABLE ENERGY TECHNOLOGIES

1.   Wilson Clark, Energy for Survival, ANCHOR PRESS / DOUBLEDAY
        (New York, New York), p. 513, 1974.

2.   Ben Kocivar, "Tornado Turbine," POPULAR SCIENCE, Vol. 210, No. 1,
        p. 1 (Cover) and pp. 78-80, January 1977.

3.   William E. Heronemus (College of Engineering, University of
        Massachusetts at Amherst), "Using Two Renewables," OCEANUS
        (Woods Hole Oceanographic Institution), Vol. 17, p. 24, Summer 1974.

4.   Clarence Zener (Carnegie Mellon University), "Solar Sea Power,"
        PHYSICS TODAY (American Institute of Physics), Vol. 26, No. 1, pp.
        48-53, January 1973.

5.   Donald F. Othmer and Oswald A. Roels, "Power, Fresh Water, and
        Food from Cold, Deep Sea Water," SCIENCE, Vol. 182, No. 4108, pp.
        121-125, October 12, 1973.

6.   W. H. Avery, D. Richards and G. L. Dugger (Applied Physics
        Laboratory, Johns Hopkins University), "Hydrogen Generation by
        OTEC Electrolysis, and Economical Energy Transfer to World Markets
        Via Ammonia and Methanol," INTERNATIONAL JOURNAL OF
        HYDROGEN ENERGY, Vol. 10, No. 11, September 1985.

7.   Donald F. Othmer, "Power, Fresh Water, and Food from the Sea,"
        MECHANICAL ENGINEERING, Vol. 98, No. 9, pp. 27-34, September
        1976.

8.   Robert L. Pons (Ford Aerospace and Communications Corporation,
        Newport Beach, California), "Optimization of a Point-Focusing
        Distributed Receiver Solar Thermal Electric System," JOURNAL OF
        SOLAR ENERGY ENGINEERING, (Paper number 79-WA/Sol-11),
        Vol. 102, No. 4, pp. 272-280, November 1980.

9.   Jim Schefter, "Solar power cheaper than coal, oil, gas," POPULAR
      SCIENCE, Vol. 226, No. 2, pp. 77-79, February 1985.

10.   Graham Walker, "The Stirling Engine," SCIENTIFIC AMERICAN, Vol.
      229, No. 2, pp. 80-87, August 1973.

11.   Andy Ross, Stirling Cycle Engines, SOLAR ENGINES (Phoenix,
      Arizona), p. 31, 1981.

12.   "Annual Technical Report," (DOE/JPL-1060-51) This report was pre-
      pared for the U.S. DEPARTMENT OF ENERGY through an agree-
      ment with NASA and the JET PROPULSION LABORATORY
      (Pasadena, California), and NASA LEWIS RESEARCH CENTER
      (Cleveland, Ohio), March 1982.

13.   George R. Dochat (Mechanical Technology Inc, Latham, New York),
      "Development of a Small, Free-Piston Stirling Engine, Linear-
      Alternator System for Solar Thermal Electric Power Applications,"
      SOCIETY OF AUTOMOTIVE ENGINEERS, INC. (SAE), Paper No.
      810457, Reference Vol. 90, p. 85, International Congress and
      Exposition, held in Cobo Hall, Detroit, Michigan, February 23-27,
      1981.

14.   E. F. Lindsley, "Solar Stirling Engine," POPULAR SCIENCE, Vol. 212,
      No. 6, p. 74, June 1978.

15.   Glendon M. Benson, Ronald J. Vincent and William D. Rifkin, "An
      Advanced 15 kW Solar Powered Free-Piston Stirling Engine," 15TH
      INTERSOCIETY ENERGY CONVERSION ENGINEERING CONFER-
      ENCE, (paper number 809-414), pp. 2051-2056, held in Seattle,
      Washington, August 1980 (Published by the American Institute of
      Aeronautics and Astronautics, New York, New York).

16.   M. A. Liepa and A. Borhan, "High-temperature Steam Electrolysis:
      Technical and Economic Evaluation of Alternative Process Designs,"
      INTERNATIONAL JOURNAL OF HYDROGEN ENERGY, Vol. 11, No.
      7, pp. 435-442, 1986.

17.   J. F. Britt, C. W. Schulte and H. L. Davey, "Heliostat Production
      Evaluation and Cost Analysis," GENERAL MOTORS (GM) TECHNI-
      CAL CENTER (Warren, Michigan), December 1979. This report was
      prepared by GM under sub-contract No. XL-9-8052-1 for the Solar
      Energy Research Institute (SERI) in Golden, Colorado.

18.   Clemens P. Work and Kenneth R. Sheets, "Behind talk of a new wave
      of oil mergers," U.S. NEWS & WORLD REPORT, Vol. 96, No. 3, p. 59,
      January 23, 1984.

CHAPTER 7:
RENEWABLE ENERGY RESOURCES

1. "International Energy Annual 1998," prepared and published by the U. S. ENERGY INFORMATION ADMINISTRATION, for the U.S. Department of Energy, Washington, D.C.

2. Wilson Clark, Energy for Survival, ANCHOR PRESS - DOUBLEDAY (New York, New York), p. 515, 1974.

3. Clarence Zener, "Solar Sea Power," PHYSICS TODAY, Vol. 26, No. 1, p. 52, January 1973.

4. William H. Avery (Applied Physics Laboratory, Johns Hopkins University), "Ocean Thermal Energy -- Status and Prospects," MTS JOURNAL, Vol. 12, No. 2, p. 52, September 1977.

5. G. Daniel Brewer (Lockheed Aircraft Company), "Cargo-Carrying Airline proposed for Liquid Hydrogen Fuel Development," ICAO BULLETIN (International Civil Aviation) February 1979.

6. Dr. M. King Hubbert, "The Energy Resources of the Earth," SCIENTIFIC AMERICAN, Vol. 224, No. 3, pp. 66-70, September 1971.

7. Arthur F. Pillsbury (former chairman of the Department of Irrigation, University of California at Los Angeles), "The Salinity of Rivers," SCIENTIFIC AMERICAN, Vol. 245, No. 1, pp. 54-65, July 1981.

CHAPTER 8:
UTOPIA: FROM HERE TO ETERNITY

1. Robert R. Birge, "Protein-Based Computers," SCIENTIFIC AMERICAN, March 1995, page 90.

2. Jonathan B. Tucker, "Biochips: Can Molecules Compute," HIGH TECHNOLOGY, Vol. 4, No. 2, pp. 36-47, Feb. 1984.

3. Mark A. Reed and Jams M. Tour, "Computing with Molecules," SCIENTIFIC AMERICAN, Vol. 282, June 2000.

4. David Dressler and Huntington Potter, "Discovering Enzymes," SCIENTIFIC AMERICAN LIBRARY, New York, New York, 1991.

5. John Rennie, "Immortal's Enzyme, SCIENTIFIC AMERICAN, July, 1994, page 14.

6. JJoannie S. Fischer, "The Cells of Immortality," U.S. NEWS & WORLD REPORT, March 20, 2000.

7.   Jared Sandberg, "The Next Big Blue Thing," NEWSWEEK, December 13, 1999, page 83.

8.   Michael Hirsh, "Computerized immortality isn't too far off," ASSOCIAT-ED PRESS, reprinted in the ARIZONA REPUBLIC, p. AA12, June 14, 1987.

9.   Robert A. Weinberg, "The Molecules of Life," SCIENTIFIC AMERI-CAN, Vol. 253, No. 4, pp. 48-57, October 1985.

10.   Gary Felsenfeld, "DNA," SCIENTIFIC AMERICAN, Vol. 253, No. 4, pp. 58-67, October 1985.

11.   Walter J. Ghring, "The Molecular Basis of Development," SCIENTIFIC AMERICAN, Vol. 253, No. 4,, October 1985.

12.   Robert Jastrow, "Superbrain, Our Brain's Successor," SCIENCE DIGEST, Vol. 89, No. 2, p. 58, March 1981.

13.   K. Eric Drexler, "Engines of Creation," ANCHOR PRESS / DOUBLE-DAY, (Garden City, New York) 1986.

14.   Peter C. Hinkle and Richard E. McCarty, "How Cells Make ATP," SCI-ENTIFIC AMERICAN, Vol. 238, No. 3, p. 106, March 1978.

15.   "Science: Life's First Building Block: Made of Clay?" NEWSWEEK magazine, April 15, 1985.

16.   Georges Millot, "Clay," SCIENTIFIC AMERICAN, Vol. 240, No. 4, April 1979.

17.   Carl R. Woese, "Archaebacteria," SCIENTIFIC AMERICAN, Vol. 244, No. 6, p. 100, June 1981.

18.   Wilhelm Pfeffer, Untersuch. Bot. Institute Tubingen (Germany), Vol 2, No. 582, 1888.

19.   Julius Adler and Wung-Wai Tso, "Decision-Making in Bacteria: Chemotactic Response of Escherichia coli to Conflicting Stimuli," SCI-ENCE, Vol. 184, No. 4143, June 21, 1974.

20.   Paul Pietsch, "The Mind of a Microbe," SCIENCE DIGEST, pp. 69-103, October 1983.

21.   Paul Pietsch, "Shufflebrain: The Quest for the Hologramic Mind," HOUGHTON MIFFLIN (Boston), 1981.

22.   Sorin Sonea, "The Global Organism: A New View of Bacteria," The Sciences, published by THE NEW YOUK ACADEMY OF SCIENCES, July/August 1988.

23. James E. Lovelock, "Gaia: A New Look at Life on Earth," OXFORD UNIVERSITY PRESS (New York), 1979.

24. John Groom, "Goddess of the Earth," NOVA (PBS Science Documentary produced by the WGBH EDUCATIONAL FOUNDATION (Box 322, Boston, MS 02134), 1986.

25. August Kekule, "Kekule's Dream," (Berichte der Deutschen Chemischen Gesellschaft, 1890). Translated by Dr. P. Theodor Benfey; from JOURNAL OF CHEMICAL EDUCATION, Vol. 35, p. 21, 1958.

26. James A. Shapiro, "Bacteria as Multicellular Organisms," SCIENTIFIC AMERICAN, Vol. 258, No. 6, pp. 82-89, June 1988.

27. Charles W. Petit and Laura Tangley, "The Invisible Emperors," U.S. NEWS & WORLD REPORT, November 8, 1999

28. Irvin Block, "The Cell: The Worlds Within You," SCIENCE DIGEST, pp. 49-50, September/October 1980.

29. "Mapping the Genes, Inside and Out," INSIGHT magazine, p. 54, May 11, 1987.

30. William S. Weed, "How about a little Viagra for your Memory?" DISCOVER, June 2000.

31. "Few dozen genes rule old age, sudyt says," WASHINGTON POST, page A6, March 31, 2000.

32. Thomas Hayden, "A Genome Milestone," NEWSWEEK, page 51, July 3, 2000

33. James B. Angell, Stephen C. Terry and Phillip W. Barth, "Silicon Micromechanical Devices," SCIENTIFIC AMERICAN, Vol. 248, No. 4, p. 44, April 1983.

34. "Tiny Robot Being Designed to Move the Body, do Surgery," THE ARIZONA REPUBLIC, p. A3, February 16, 1989.

35. "A Journey into Smallness," COMPRESSED AIR, Vol. 92, No. 4, pp. 8-13, April 1987.

36. O. Marti, H.O. Ribi, B. Drake, T.R. Albrecht, C.F. Quate, P. K.Hansma, "Atomic Force Microscopy of an Organic Monolayer," SCIENCE, Vol. 239, p. 50, January 1, 1988. (Note: O. Marti, B. Drake and K. Hansma are at the Department of Physics, University of California, Santa Barbara, 93106; H. Ribi is at the Department of Cell Biology at Stanford University (Stanford, CA) and T. Albrecht and C. Quate are at the Edward L. Ginzton Laboratory and the W. W. Hanson Laboratories of Physics, Stanford University).

37. Vladilen S. Letokhov, "Detecting Individual Atoms and Molecules with Lasers," SCIENTIFIC AMERICAN, Vol. 259, No. 3, pp. 54-59, Sept. 1988.

38. Robert C.W. Ettinger, "The Prospect of Immortality," DOUBLE DAY AND COMPANY, (Garden City, NY) 1964.

39. THE CRYONICS INSTITUTE, 24443 Roanoke Oak Park, Michigan 48237 (313) 548-9549

40. THE IMMORTALIST SOCIETY, 24041 Stratford Oak Park, Michigan 48237 (313) 548-9549

41. "Signing Up Made Simple: A Guide to Making Cryonics Arrangements with Alcor," ALCOR LIFE EXTENSION FOUNDATION, pp. 104-107, 1987.

42. Mortimer Mishkin and Tim Appenzeller, "The Anatomy of Memory," SCIENTIFIC AMERICAN, Vol. 256, No. 6, p. 89, June 1987.

43. Alan Herringrton, THE IMMORTALIST, Vol. 19, No. 5, pp. 4-6, May 1988.

44. Peter Aleshire, "Patients Are Chilled for Brain Operations," THE ARIZONA REPUBLIC, p. B1, September 22, 1988.

45. Bernard Strehler, "The Understanding and Control of the Aging Process," Challenging Biological Problems, Edited by John Behnke, OXFORD UNIVERSITY PRESS (New York) 1972.

46. Leonard Hayflick, "Theories of Biological Aging," EXPERIMENTAL GERONTOLOGY, Vol. 20, pp. 145-159, 1985.

47. Roy L. Walford, M.D., "Maximum Life Span," W. W. NORTON, 1983.

48. Roy L. Walford, M.D., "The 120 Year Diet," SIMON & SCHUSTER (New York, NY), 1986.

49. Kathleen Stein, "Supergene: Decoding the Secrets of Immortality," OMNI, Vol 3, No. 3, p. 126, Dec., 1980.

50. "The Future of Medicine," TIME MAGAZINE, January 11, 1999.

CHAPTER 9:
CONCLUSIONS

1. Philip Shabecoff, "White House dilutes 'greenhouse' warning," THE NEW YORK TIMES release, published in the ARIZONA REPUBLIC, p. A4, May 8, 1989.

2. Peter Hoffmann, "Hydrogen gets renewed attention as fuel of the future," THE WASHINGTON POST release, published in THE GRAND RAPIDS PRESS, p. D5, October 11, 1987.

3. Anne Q. Hoy, "U.S. smog to persist, report says," THE ARIZONA REPUBLIC, pp. A1-A6, July 18, 1989.

4. Thomas A. Sancton, "What on Earth Are We Doing?" TIME INC. MAGAZINES, Vol. 133, No. 1, p. 30, January 2, 1989.

5. "Northern Ozone Layer 60% Lost," Knight Ridder Newspapers, published in the ARIZONA REPUBLIC, page A-12, Sunday, April 9, 2000.

6. Steve Wilson, "It's time leaders were honest about gas prices," ARIZONA REPUBLIC, Page A2, June 24, 2000.

INDEX

A New Bacteriology, by Sorin Sonea 281
Acid deposition 7, 90
Acid fallout 8
Acid fog 8
Adam 269
Adler, Julius 277
Advanced Strategic Computing
  Initiative White (ANCI White) 256
AEG Micon 196
Aerobic 103
aerosol spray cans 14, 21
age of exponentials 47
age of microorganisms 98
agng 254, 259, 264, 303,
  308, 334, 336
agricultural systems 3, 31, 34, 315-317, 335
AIDS virus afflictions 45
air conditioning systems 14
Air Force Flight Dynamics Laboratory 127
Air Force Liquid Hydrogen-Fueled B-57 135
air Pollution 29, 322
Air Products & Chemicals 106
Airbus 322
algae 12
Alkon, David L. 98
Allen, Woody 307
alternative fuels 318
American Academy of Science 122-123
American Cancer Society 31
American Institute of Aeronautics and
  Astronautics 185
American Journal of Physics 39
American Petroleum Institute 45
American Water Works Association 30
American Wind Energy Association 199, 239
amino acids 97, 258, 270-271
Amundson, Robert 27
Antarctic ozone levels 20
Appleby, John 318
Arab oil embargo 58, 67, 152, 182, 222-224
Aral 329
arcologies 147
Arcology 147
Argonne National Laboratory 84
arithmetic vs. exponential growth 38
Arizona Corporation Commission 198
Arizona Public Service Company 172
Arizona Republic 331
ASCI Blue Pacific 256
AT&T laboratories 293
Atlas of the United States 231
Atmospheric gain of carbon dioxide 9
Atmospheric gases 13
Atom 292, 295-296
Atomic Force Microscope (AFM) 295
Atoms 15, 104, 156
Automotive Stirling engine 175
Avery, W. H. 218, 240
Aviation kerosene 127
Aviram, Arieh 258
Backhaus 188

Backhaus, Scott 188
Bacteria 12, 100, 152, 266, 268,
  274-275, 277, 282
Bacterial fossils 101, 269
Bain, Addison 125
Barrows Neurological Institute 301
Bartlett, Albert A. 38, 64, 66
Beale, William T. 183
Benson, Glendon 185
benzene 27
benzene emissions 7
Big Bang 94, 96, 268
Billings Energy Corporation 109, 123
Billings, Roger 120-121
Binnig, Gerd 294
biochemical molecules 9
Biochip 258-259, 264, 290
Biocybernetic; 292, 298, 306, 309, 336
  era 299
  evolution 264
  organism 263, 283
  utopia 309
biological effects of light 21-26
biological impacts of sunlight 21-26
biological life support systems 6, 154
biological mechanisms of aging 254, 259, 264,
  303, 305, 308, 334, 336
biological pathways for light 22, 23, 24
biological transition to
  renewable resources 253, 334
biomass 61, 172, 187
biomass burner 187
biomass resources 4
bioStirling Systems 187
biotechnology 46, 254, 266, 291, 299
bisexual reproduction 305
Block, Irving 287
Blue Gene 256-257
BMW 111, 115, 116, 118,
  201, 324, 327, 329
Boeing 119, 136, 173, 181,
  187, 194, 202, 322
Boeing/SES Solar Test Site 170
Bonneville Power Administration 330
BP/Amoco 323
Bratt, Christer 181
Brattle Group 335
Braun, Werner Von 136
breeder nuclear reactors 73
Brewer, G. Daniel 129
Brilliant, Ashleigh 307
British Antarctic Survey 20
British Interplanetary Society 143
British Petroleum 244, 316, 329
British Thermal Unit (Btu) 61-62, 151, 159,
  187, 201, 229, 241,
  325, 331, 332
Broecker, Wallace 11
Brookhaven National Laboratory 109, 246
Brower, Kenneth and David 89
building blocks of life 101
bulldozer culture 33
Bush Administration 222
Bush, Vannovar 195
Bussard, R.W. 143
Cairns-Smith, Graham 270
calcium 23, 25
California Institute of Technology 8, 88, 280
Caltech 258

cancer cells 16, 27, 31, 260, 274
carbon 96-97, 103, 151, 156, 270, 322
carbon atom 11
carbon cycle 12
Carbon Cycle Corporation 187
carbon deposits 119, 174
carbon dioxide 9, 12-13, 32, 45, 62
carbon dioxide emissions 10
Carnegie-Mellon University 104, 218, 240, 266, 282
Carter Administration 211, 222, 331
Cavendish, Henry 104
Columbia University 11
central receiver system 190-191, 194
centralized photovoltaic array 163
cents per kilowatt hour 160
Ceperley, Peter 188
Chadwick, Henry 311
chemical contamination 7, 21, 27, 31
chemical reactions 11
Chernobyl accident 83
chlorine 15
chlorofluorocarbon (CFC) molecules 9, 14, 16
CFC industries 21
chlorophyll 102
chromosomes 260
Chrysler 293
city of the cell 267, 285-286
civilization 2, 6, 31
clay 271-272
Clean Air Act 30
CleanEnergy Exhibition 327
climate change 10, 29
climatic "stable-state" 13
Clinton Administration 222, 316, 318
coal; 107
    plants 62
    production 61
    gasification 151
Cocking, Dennis 30
cold fusion 86, 316
Cold Spring Harbor Laboratory 261-262
Columbia University 11, 219, 264
combustion of fossil fuels 13
common ancestor 269, 273
Commoner, Barry 89
Comparative Analysis of Solar Technologies 194
computers 46, 256-257, 259, 264-265, 291, 294, 303, 311
conclusions 55, 71, 90, 153, 225, 252
consciousness 277, 308
continental shelves 13
cool white fluorescent lamps 24-26
Cornell University 47
cosine losses 190
Cosmos, by Carl Sagan 47, 97, 144, 156, 254
Coyne, Lelia 270
crankcase oil 27
creationists 267
Crick, Francis 287
crop losses 32
crude oil reserves 63
Cryenco Inc 189
Cryobiology 299
cryogenic fuel storage tanks 117
cryogenic preservation 300
cryogenic tanker 236
cryogenic ark 299

cryogenic suspension 298, 309
Cummins Diesel Engine Company 186
cybernation 264
cybernetic 294
cybernetics 264
Cytochrome enzyme 261
D'Arsonval 218
Daimler-Benz 109
DamilerChyrsler 324
Davis, Noel 313
death; 25, 31, 266-267, 269, 271, 301, 303-311, 314
    defining 301
    of forest trees 32
    of the oceans 312
Deep Blue 256-257
defective genes 289, 292
deforestation 13, 31-32, 45, 312
Denneau, Monty 257
Deoxyribonucleic acid (DNA) 15
Department of Defense 123, 127
Department of Energy 182
depletion of the Earth's stratospheric ozone 3
dermatologists 25
desalinated seawater 247, 252
desertification 45
designer genes 2, 254, 263, 297
designer gene era 263, 297
destruction of forests 9
development of green plants 102
Dickerson, Richard E. 261
Diesel engines 174, 176-177
Diesel, Rudolf 174
Diesel-cycle engine 174
dinosaurs 13
Discover magazine 188, 262
Dispatchability 200
DNA 97, 107, 260, 266, 269-271, 273-275, 286-287, 289, 299, 303
DNA Helix 255
Drexler, K. Eric 265, 290-291, 298, 300
driftnetting 206
droughts 9, 32, 312
early Solar Stirling Systems 166
early Stirling engines 182
Earth; 1, 10, 93
    atmosphere 13, 32
    biological life support systems 4
    biosphere 33
    climate 32
    environmental life-support systems 2
    human population 6
    ocean ecosystems 206
    ozone shield 21
    protective ozone shield 16
earthquake 33
ecological disruptions 33
economic considerations 57, 220
economic growth 65
economically recoverable energy reserves 68
education 54, 58
education: problems and solutions 50
Edwards Air Force Base 165
Ericsson, John 166-167
Einstein, Albert 15, 73
electric battery weight 111
electric vehicles 111
electric waves 16

electricity   4, 62, 111, 153, 198, 223, 245
electricity costs from natural gas   70
electrolysis of water   107
electrolytic hydrogen   152
electrolyzer   241
electromagnetic radiation   22, 25, 93, 162
electron-transferring proteins   258
electrons;   11, 15, 94, 96, 124, 162, 270
  energy levels   11, 15
elements of the Earth   97
emissions   29
energy;
  and environmental problems   34
  conservation   34
  crisis   58
  economics   160
  from wastewater treatment plants   172
  fundamentals   61
  wavelengths   15
Engines of Creation, by K. Eric Drexler   265
EnronWind Corporation   196
entertainment   52
environment   75
environmental costs   3
environmental damage   1
environmental dislocations   34
environmental problems   311
Environmental Protection Agency   21, 27, 198
Environmental systems collapse   33
enzymes   97, 101, 258, 266, 279, 286, 289, 304
enzyme engineering   289
Erren, Rudolf A.   119-120
ethanol   149
ethical Questions   305
ethylene dibromide   27
Ettinger, Robert C. W.   299
eukaryote   285
Eve   269
evolution   98, 102, 270, 275
evolution of life on Earth   98
exponential;
  growth   15, 22, 30, 31, 33-34, 37-55, 222, 254, 291, 302, 311-312, 334, 336
  concept of 11:59   42
  destruction of the remaining wilderness and wildlife habitats   68
  doubling times   40
  expiration time   65
  forces   47
  growth of the human population   2, 44
  icebergs   45
  increases in wind speed   202
  increases of nanobes   101
  savings   42
  diminishing natural gas reserves   70
exposure to solar radiation   26
external combustion   172, 174
external costs   161, 220, 224, 323
Exxon   316
Fair Accounting Act   253, 321-322, 324, 330
fair accounting system   220, 224
Farman, J.C.   20
Federal Power Commission   195
fermentation   101
Feynkman, Richard   258

Fickel, Hans-Christian   325
Fleischmann, Martin   86
Fluor Daniel Corporation   330
fluorescent lamps   25, 26
food and water shortages   31
food chains   3
food production systems   31, 315-317, 335
Ford Motor Comapny   111, 175, 182, 201, 292, 325
Foresight Institute   265
forest death   8
forest fires   32
forests   8, 31
fossil fuels;   3, 4, 8, 12, 13, 57, 61-62, 64, 65, 68, 85, 90, 93, 105, 159, 161, 166, 186, 198, 200, 222, 225, 229, 230, 232, 311
  reserves   65-66
  power plants   83
fossils   269
free market forces   155
free-piston Stirling engine   182-184
Freon™   14
fresh water aquifers   247
freshwater pipelines   246
Front-end of the nuclear fuel cycle   87
Frontline   53
Fuel cells   111, 117, 327
Fujimasa, Iwao   294
Full spectrum lamps   24, 26
Fungi   12
fusion research and developmen   86
Future Shock   48
Gaia: A New Look at Life on Earth   13
  by James Lovelock   282
Galaxies   1, 156
gas lines   67, 332
gasoline;   4, 27, 29-30, 107, 118-119, 125, 149, 200, 323, 332
  and diesel fuel storage tanks   7
  fueled engines   174
Gates, Bill   319
gene chips   303
gene therapy   255
General Dynamics   235-236
General Motors   182, 191, 292, 325
genes   254, 297
genetic code   301
genetic engineering   266, 276, 305
geologic carbon cycle   12
geologic plates   33
George Mason University   188
geothermal energy   61, 150
Glaciers   32
global;
  acid deposition   7
  atmospheric and weather systems   31
  biomass resource   187
  climatic change   31-32
  environmental problems   33
  food production systems   6
  living system   12
  Solar Resources   234
  systems collapse   6
  warming   9, 21
Gottingen University   102
Government Regulation   319
gravity   93, 96, 143

greenhouse gases 3, 7, 9, 14, 32, 90, 224
greenhouse warming 31
Greider, Carol W. 261
groundwater supplies 198
Grove, Sir William 111
Grumman 207, 216
Gulf Oil 243-244
Haber Institute 294
Haldane, J.B.S. 195
Hansen, James 10
Harrenstien, Howard 150
Harrington, Alan 308
Harvard University 26, 299
Hawkins, Willis 154
heat wave 31
heliostats 190-191
Heronemus, William E. 84, 202, 204, 206, 210, 219
Hertz, Heinrich 162
high school students 52
high-speed electron 89
high-temperature electrolysis 246
Hindenburg 124-127
Hinkle, Peter C. 100
Hitachi 294
Hoffmann, Peter 120, 154
holograms 278
holographic storage systems 279
Homo Immortalis 336
Horgan, John 86
hormone 22
Horn, Paul 257
Hubbert, M. King 64-65, 224
human;
  genome project 288
  health 21
  population 13
  population growth 41, 44
hurricanes 32
Hurst, Samuel 296
Hutton, James 282
hydrocarbon fuel spills 28
hydroelectric power plants 61, 229, 251
Hydrogen 4, 7, 12, 93, 97, 102-105 129, 145, 150, 153, 156, 159, 166, 180, 195, 201, 203, 206, 223, 226, 233, 270, 279, 316-317, 322, 326-327, 336
  aircraft applications 129
  atom 94, 96
  atoms 157
  bomb 124
  cryogenic tanker 23
  economy 3, 224
  energy content in water 243
  energy system 3, 155
  engines 119
  explosions 127
  fire 129
  flammability 125
  for Japan 237
  from water 14, 243
  fuel for
    BMW Engine 112
    Coleman Stove 123
    Mercedes 110
    Saturn 5 moon rocket 138

stars 96
homestead 122
hydrides 109
ions 7
nuclei 15
production 151
production board 317
refueling station 115
renewable fuel cycle 112
residential delivery system 110
safety 124
scoop 143
shuffling 104
storage systems 108
scoop space colony 146
tree 106
hypersonic;
  aerospacecraft 139
  aircraft cutaway 141
hyperviolent films 52
IBM 256, 258, 292, 294-295
Ice Age 12
icebergs 32, 38, 46
immortal bird 4
immortalizing enzyme 260
immortality 307
immune and endocrine system 22
implementation lead-times 241
Indiana University 278
indirect solar resources 61
industrial revolution 13
industrial transition to renewable resources 6
inflation 59-60
information explosion 49
information gap 223, 317
infrared;
  heat from the Sun 9
  radiation 9
Institute of Spectroscopy 296
insurance industry statistics 32
Integral Fast Reactor 84
internal combustion 182
International Association for Hydrogen Energy 4, 155, 221, 223, 244
International Journal of Hydrogen Energy 155
interrelationships 57
Interstate highway system 5
Iran 63
Iraq 63
Iron-titanium hydrogen hydrides 109
Jastrow. Robert 264
Jet Propulsion Laboratory (JPL) 165, 166
John Deere 186
Johns Hopkins University 216-218, 240, 258, 289
Kasparov, Garry 256, 257
Kawasaki Heavy Industries 235, 237
Kekule's Dream, by August Kekule 284
kerosene 129
kilowatt-hour (kWh) 229
kinematic Stirling engines 182
Kingdom of Saudi Arabia 233
Kockums 168, 174-181
Koshland, Daniel 278
Kuwait 63
Kyoto Protocols 10
La Cour, Poul 195
land;

in US needed for displacing
fossil fuels with solar technologies   230, 232
in Saudi Arabia needed for displacing
oil exports with solar technologies   233
landfills   152
Lavi, Abraham   104
Lavoisier, Antoine   104
Law of the Cube   202
Leaking fuel storage tanks   27
Leisure time   53
Lemley, Brad   188
Lemmings   34
Lerner, Richard   304
Letokhov, Vladilen S.   296
Lewis, James B.   299
Life-Cycle Considerations   242
Lifeboat Agricultural Systems   313
light   162, 296
    lquantity vs. quality   25
limestone   11
Linde   118, 329
Line-focus systems   192
Lipetz, Philip   306
liquid hydrogen;   111, 113, 118, 129,
    139, 151, 237, 329
    Cooling System   142
    fuel for   108
    airport   130-131
    Boeing 747   132
    Buick   114
    cars   118
    launch Vehicle   136
    Lockheed Freighter   134
    Space Shuttle   135
    fuel Storage   130, 236
    Lockheed Aircraft Design   131
    pipelines   238
    safety   113
    self-serve refueling station   114
    storage tank   116
    tank   124
    tank weight   111
    tanker trucks   238
    wing tanks   132
liquid sodium   84
Lockheed   113, 119, 129-130, 136, 154,
    211-212, 214-215, 242, 322
Lopez, Charles   194
Los Alamos National Laboratory   114, 117, 188
Lovelock, James E.   12, 282
Lucent Technologies   279
Lung cancer in women   21
Luz International Limited   192-193
M.I.T.   68
Macro-engineering   5
Mahlman, Jerry   10
malignant melanoma   20
mammals   1
MAN   329
Manabe, Syukuro   10
Manhattan Project   5, 50, 324
Mars   14, 20
Marsano, Doug   30
mass extinction   1
mass-production of renewable
    energy technologies   220
Massachusetts Institute of Technology   26, 147,
    210, 265, 313
Matsushita Research Laboratories   292

Mazda   325
McDonnell Douglas   140, 142,168-170,
    173, 194
McDonnell Douglas Aerospace Plane   140
McDonnell Douglas solar R&D   169
Mechanical Technologies Inc.   184
Media   316
Megawatt   229
Memory proteins   262
Methane   9
Methanol,   149
Methyl tertiary butyl ether (MTBE)   27
Microbial memory   279
Microbes   13, 97, 267, 280
Microbial Evolution   276
Microbial food-chains   152
Microbial Hydrogen-fueled Engine   100
Microbial Memory   277
Microbial mind   278
Microbial superorganism   13
Microbial world   274
Micromachined Gears   293
Microorganisms   9
Microrobot   294
Microsensor Technology Inc.   292
Microtechnology   292
Middle East   61, 63, 67, 88, 156, 223
Milberg   326
Milberg, Jocahim   325
Millot   272
Millot, Georges   271
Mind Children, by Hans Moravec   264
Ministry of International Trade
    and Industry (MITI)   186, 237, 294
Mion, Pierre   173
MIT   66, 265
MITI   320
Mitochondria   287
Mobil Oil Corporation   66
Mobile Stirling PCU   181
Modularized energy systems   221
Mohnen, Volker   8
Molar teeth   24
Molds   12
molecular;
    biology   46, 254, 259, 266,
        269, 280, 299, 304,
        311, 334
    chemistry   15
    damage   16
    mechanisms of memory   98
    medicine   299, 305
    scale;
        rotary engine   100
        technologies   2
    molecules of memory   262, 299, 302
Moller Aerobot   148
Moller International   148
Moller, Paul   148
Momentum   54
Moravec, Hans   253, 264
Moscow News   83
Mother Nature   44
MTBE   30
multi-array wind turbines   202-206
multi-array offshore wind system   202-204
municipal sewage   152
Murray, Allen   66
nanobes;   97-98, 266, 267-268, 280, 285

nanobe evolution 266
nanobes molecular machines 102
nanobial ancestor 280
nanobial civilization 271
nanobial consciousness 285
nanobial enzymes 98-99
nanobial evolution 97, 267
nanobial organisms 104
nanobial origins 270
nanobial proteins 101, 269
nanobial reindustrialization effort 102
nanocomputers 289-290
nanometers 22
nanoorganisms 97
nanoscopes 269, 294
nanotechnology 265, 291-292, 294, 299, 305, 307
nanotechnology era 297
nanotechnology of transhydrogenation 101
NASA 10, 20-21, 111, 125, 127, 135-137, 175, 177, 182, 186, 196, 201, 270, 282, 322, 335
NASA Stirling PCU 184
National Aerospace Plane 139
National Geographic Society 205
National Hydrogen Association 244
National Institute of Neurological and Communicative Disorders and Stroke 98
National Institute on Aging 260
National Oceanic and Atmospheric Administration's Geophysical Fluid Dynamics Laboratory 10
National Renewable Energy Laboratory 187
National Science Foundation 258, 294
natural gas; 4, 61-62, 70, 107, 149, 151, 189, 198, 201, 235
  electricity costs 162
  liquids 61
  well Productivity 70
natural outdoor environment 25
natural outdoor sunlight 21
NAWAPA 247-248, 251
NAWAPA Western Region 249
neuroendocrine organs 23
neurons 98
neutrons 11, 15, 75
New York Academy of Sciences 281
New York Times 22
New York University Medical Center 20
Newsweek 243, 257
nitrogen 9, 103, 270
nitrous oxides 7
Nixon 331
Nixon Administration 222
Nonrenewable sources of energy 61
Nova 53, 282
NovaSensor 292
nuclear energy 4, 61, 73, 86, 93, 107, 159-160, 186, 195, 311, 220, 222, 224, 229
  back end of the nuclear fuel cycle 76
  chain-reaction 76
  control rods 76
  decommissioning of nuclear reactors 76
  economics 87
  electricity costs 160
  electricity production 76
  explosions 83

facilities 186
national sacrifice zones 82
nickel-59 76
fission 73, 93
fusion 86, 93
meltdown 76
reactor maintenance 83
safety 83
war 1, 38
waste;
  package designs 80
  problem 78
  spreading problem 75-81
  storage 78, 81
  weapons 76, 82
operating nuclear power plants in the U.S. 74
spent fuel rods 77
Yucca Mountain nuclear waste storage site in Nevada 78, 82
nucleus of the cell 15, 287
O'Neill, Gerard K. 142, 144, 145
Oak Ridge National Laboratory 296
oblivion 2, 33, 49, 253, 298, 311, 315, 336
oblivion scenario 6, 16
ocean thermal energy conversion (OTEC) 61, 150, 164, 212
Ocean Wind Energy Systems 202
oceans; 11, 202, 206
  on Mars 14
octane 27
offshore accidents 28
offshore wind system 202-203, 205-206
Ogden, Joan 232
Ohio State University 306
Ohio University 183
Oil; 45, 57, 61-62, 65, 161, 232, 333
Oil and Gas Journal 61
oil companies 221, 244, 317
oil economy 224
oil reserves 224
oil spills 7, 28, 224, 239
oil surplus 67
Olson, Arthur J. 99
Omni 89
Ordin, Paul 127
Organic molecules 157
Organization of Petroleum Exporting Countries (OPEC) 60
OTEC; 211, 218-219, 224-225, 230, 241
  aquaculture 219
  capital costs 211
  design by Grumman 216
  design by Johns Hopkins University 217
  design by TRW 217
  fresh water production 219
  Lockheed Cutawa 213
  Operation 214
  Resources 240
  Sea Water Desalination 251
Other Alternative Fuels 149
Othmer 63
Otto cycle 174, 182
Otto, Nikolaus 174
over-fishing 206
overpopulation 297
oxygen 9, 14, 96-97, 103, 145, 151, 244, 270

oxygenated fuels 30
ozone 14
Parsons Engineering Company 248
Pease, R. Fabian 295
Pesticides 3
Petersen, Kurt 292
Petit, Charles 274
Pfeffer, Wilhelm 277
pH scale 7
Phoenix 4
Phoenix bird 96
photobiology 22
photobiology study 26
photochemical smog 27
photoelectric effect 162
photon 16
photons 15, 22, 93
photosynthesis 102
photosynthetic microorganisms 14
photosynthetic processes 153
photovoltaic;
  cells 61, 161-164, 185, 190,
       194, 225, 230, 232
  costs 162
  system efficiencies 162
Physical Geography: Earth Systems 234
Physics Today 142
Phytofarms of America, Inc. 313-314
Pietsch, Paul 280
Pillsbury, Arthur F 250
plankton 12
plasmids 281
plutonium 85
plutonium economy 73, 85
point-focus concentrator systems 61
point-focus solar concentrator system 164
polar clouds 335
polar icecaps 32
political action 50
political priorities 317
Polytechnic Institute of New York 63
Pons, Stanley 86
Popular Science 57, 60, 183, 204,
       207-209, 293
positive exponentials 46
power conversion unit (PCU) 164
Power Engineering 196
power towers 190
primary cell types 285
primary vs. secondary energy sources 107
primordial hydrogen 96
primordial sea 145
primordial soup 102
Princeton University 10, 142
priorities 307, 311
projected production of the
  worlds fossil fuels 59
Prokaryotes 285
Prospect of Immortality
  by Robert C. W. Ettinger 299
Proteins 97-98, 257, 261-262,
       278, 266-268, 278, 286,
       287297, 305
  engineering 305
  folding 257, 271
  proton 11, 15, 94, 96
prototype Mini OTEC Plant 215
Public Broadcasting System 53, 156, 282
pyrolysis-gasification 152

radiation 89
radioactive isotopes 9, 75
radioactive wastes 7, 75, 76, 81, 88, 224
railroads 5
rain forests 13
Ralph M. Parsons 247
RAND Corporation 66, 68
Ratner, Mark 258
Ravin, Jack 27
Reagan 221, 222
Reduced energy consumption 34
Relative Abundance of Elements 103
Relative Fuel Densities 141
Renewable energy resources 152, 229
Renewable energy technologies 159, 199,
       220, 324
Reordering of the national and state priorities 49
Resolvable Environmental Problems 7
Resonance-ionization spectroscopy 292, 296
Ribonucleic acids (RNA) 269, 275
Ribosomes 287
Rickover, Hyman 84
Rigel, Darrel 20
Rockefeller, John D. 319
Rockwell International 279
Rohrer, Heinrich 294
Royal Academy of Sciences 294
Ruska, Ernst 294
Russia 67
Russian Academy of Sciences 296
Sack, George H. 289
Sagan, Carl 47, 97, 144, 156, 254, 336
salt accumulation in the soil 312
Salt River Project 172
Santee, Michele 335
Saturn 5 moon rockets 136
Saudi Arabia 63, 67
Saving Private Ryan 52
Scanning tunneling microscope (STM) 292, 294
Schlegel, Hans 102
Science 10, 277, 303, 335
Science Applications International
  Corporation (SAIC) 172, 185
Science Digest 287
Scientific American 8-9, 23, 26, 31, 86,
       98, 249-250, 260, 261,
       271-276, 279, 296, 323
SCR 189
Scripps Clinic 99, 267
Scripps Research Institute 304
Sea levels 32
Seaborg, Glenn T. 50
Second Law of Thermodynamics 211
Seiden, Philip 258
Senate Energy
  and Natural Resources Committee 10
Sex organs 26
Shale oil 63
Shapiro, James 274
Shaw, Robert W. 8
Shell 316, 323, 329
Shell Hydrogen 244
Shell Oil 244
Short-wave cosmic radiation 6
Shufflebrain:
  The Quest for the Hologramic Mind,
  by Paul Pietsch 278
Sick buildings 25
Siemens 185, 329

siting of wind machines 200, 202, 203
skin in humans 23
skin cancer 20, 25
Sleeper 307
soil moisture 10
solar energy; 93, 150,230,
concentrator major subassemblies 169
dish forest 173
energy received by the Earth 230
engine systems 164-165, 182, 190
land requirements 166
global radiation (SGR) 25
heliostat 190-191
hydrogen technologies 4
insolation in the U.S. 230
PCU maintenance 170
radiation 15, 22
resources 230
Stirling engine;
systems 172, 178, 179,
224, 225, 246
cutaway 178
Stirling PCU 179
Stirling power plant 173
solar trough systems 193
Solar Hydrogen: Moving Beyond Fossil Fuels
by Joan Odden and Robert Williams 232
Soleri, Paolo 147, 210
solutions 312
Sonea, Sorin 281
soul 269
South Tahoe Public Utility District, California 30
Southern California Edison 168, 191, 194
sapace colonies 145, 149
space habitats 142, 298
space shuttle 136, 139
sperm production 24, 26
Spetzler, Robert 301
spirit 269
SST program 139
St. Lawrence, Ethel 302
Stanford University 265, 280, 292-293, 295
Starship Hydrogen 298
State University of New York at Albany 8
Status of all ordered nuclear units 74
steam engines 174
Sternback, Rick 143
Stirling engines 194, 230
4-95 engine 175, 176
cryocoolers 189, 241
engine cutaway 176
engine manufacturing costs 186
free-piston Stirling engine 182-184
heater head 178, 180
heater head quadrant 180
power conversion unit (PCU) 168, 177, 182
PCU "on sun" 179
PCU on bench 181
sun motor 167
Stirling Energy Systems (SES) 168,173, 176,
178, 187, 202
Stirling Thermal Motors (STM) 172, 185, 187
Stirling, Robert 174
Stirling-cycle 189
Stirling-cycle engine 166, 174
Stirling-powered attack submarine 177
Stone, Kenneth 181
storage of the radioactive wastes 29
strategies for survival 49

Stratified-charge combustion 174
stratospheric ozone depletion 6-7, 14-15, 20,
45, 335
Strehler, Bernard 303
Strobl, Wolfgang 325
Structural Theory 285
students 52
subconscious 284-285
submandibular gland structures 24-26
Sumitomo Electric Industries 294
sun 93
sun motors 167
sun, heat output 13
sunlight 21, 25
sunlight and health 23
supercomputers 20, 256-257
superorganism 13, 280-282
superparamagnetic (SPE) effect 279
surface membranes of cells 287
Sussman, Gerald 265
Tabazadeh, Azadeh 335
Tangley, Laura 274
Tappan Appliance Company 122
Tappan Hydrogen-Fueled Gas Range 122
tar sands 63
TASHE PCU 188
teaching teachers 52
technological trees 166
Telomerase 263
Telomeres 261
temperature of the Earth 12
terrestrial solar electromagnetic radiation 25
Texaco 244, 316, 323
Texas A&M 318
Texas Tech University 59
*The Ages of Gaia: A Biography of
Our Living Earth*, by James Lovelock 13
The Battery 68-69
The Closing Circle, by Barry Commoner 89
The Forever Fuel, The Story of Hydrogen,
by Peter Hoffmann 120, 154
*The High Frontier: Human Colonies
in Space*, by Gerard K. O'Neill 142
The Immortalist, by Alan Harrington 308
The New York Times 77
The Phoenix Project 330
thermal load of nuclear waste 80
thermoacoustic Stirling engines 188-189
thermochemical water splitting 244
Third World countries 66
Thomas, Percy H. 195
Thorndike 58
Three Mile Island 75, 128
Till, Charles E. 84
Time Magazine 243, 268, 318
Time. Cells and Aging, by Bernard Strehler 303
Titanic 36-37, 44, 54, 71, 312, 315
Toffler, Alvin 48
Toigo, Jon 279
Tokyo University 294
tooth decay 24
toxic substances in gasoline 27
Toyota 294
transhydrogenation 101
Trends 33
tropospheric ozone 27
TRW 217
Tso, Wung-Wai 277
Tully, Tim 262-263

United States;
 annual energy demand 187
 coal production 62
 Congress 214, 221, 288, 319, 330
 Defense Advanced Research Projects
  Agency (DARPA) 279
 Department of Commerce 231
 Department of Energy 31, 165, 187, 211, 242
 energy consumption in 1998 229
 Energy Information Administration 61-62, 69-70,
  74, 201, 229
 energy policy 222, 331
 energy production and consumption 159
 energy production in 1998 229
 Geological Survey 45, 65, 224
 oil imports 34
 military costs in the Middle East 161
 National Oceanic and Atmospheric
  Administration's Fluid Dynamics
  Laboratory 10
 natural gas production 71
 Naval Academy 202
 Naval Research Laboratory 173, 258
 oil consumption 64
 oil production 67, 69, 241
 Senate Hearing of the Energy and Commerce
  Health and Environmental Subcommittee on
  ozone Depletion 20
 Solar Resources 231
 War Production Board 5
ultraviolet radiation 14-16, 22-25, 125
underground gasoline storage tank leaks 30
Union Carbide 238
United Arab Emerites 63
universal fuel 4
universe 1, 103, 156
Universite Louis Pasteur 271
University of Arizona 280
University of California 50, 292, 295, 303
University of California at Berkeley 278
University of California at Davis 148
University of California at Los Angeles 8, 258
University of Chicago 274
University of Colorado 39, 64
University of Glasgow 270
University of Massachusetts
 at Amherst 84, 204, 202, 218
University of Miami 150
University of Michigan 292
University of Rochester 58
University of Southampton, England 86
University of Tokyo 292
University of Utah 86
University of Washington 300
uranium; 93
 239 77
 atoms 75
 fuel enrichment facilities 87
 fuel rods 75
 reserves 90
utopia 2, 33, 49, 253,
  311, 315, 336
vacuum-jacketed cryogenic fuel lines 130
vacuum-jacketed storage tanks 117
Venus 14
Verleger, Phil 335
Vertical Vortex wind system 207
Vestas wind macnines 196, 198
Vestas 660 kW Wind Turbines 197

Vestas Power Conversion Unit 197
Veziroglu, T. Nejat 155
viruses 97, 268
Vitamin D 25
Volcanic eruptions 11
Volcanoes 8
Volvo 175, 186
Vortex 208
Vortex Arcology 210
Vortex Cutaway 209
Vortex Generator 208
W.R. Grace's Davison Chemical Company 77
War Production Board 317
Warner, Huber 260
Washington Post 303
Watanabe, Tsutomu 281
Water 4, 25, 243, 322
 options 245
 pollution 27, 30
 supplies 32
Watson, James 287
Watson, Robert 21
wavelength; 15-16, 22, 162, 296
 and energy level 16
Wayne State University 299
Weather records 31
West Valley facility 78
West Valley nuclear reprocessing plant 77
Wetherald, Richard 10
wholesale electricity costs 160
Williams, Robert 232
Wilson, Carroll 66
Wilson, Edward O. 274
Wilson, Steve 331
wind energy; 61, 150, 219,
  224, 230-231
 electricity costs 196
 conversion systems 164, 195
 resources 199, 239
 hydrogen 7
 machines 202
window glass 25
Windships 206
Woese, Carl R. 275, 276
Woods Hole Research Center 9
Woodwell, George M. 9, 31
world energy production in 1998 229
World Magazine 107
world oil discoveries 57
world oil reserves 241
World War II 50, 67, 224
Wright, Robert 268
Wurtman, Richard J. 26
Yale University 275
Yates, F. Eugene 259
Yen, James T. 207
Yin, Jerry 262
Yucca Mountain nuclear waste
 storage site 78, 82
Zener, Clarence 104, 240-241

Harry Braun
Chairman
Sustainable Partners, Inc.

About the Author

Harry W. Braun has worked as an energy analyst for the past 20 years and is Chairman and CEO of Sustainable Partners, Inc., a Phoenix-based corporation that is teamed with Stirling Energy Systems (SES) in the commercialization of state-of-the-art, and advanced solar, wind and hydrogen production systems. Mr. Braun was one of the original founders of SES, which has its engineering offices at the Boeing facility in Huntington Beach, California. Mr. Braun was instrumental in developing strategic business relationships with McDonnell Douglas (now Boeing), as well as the BMW Group, Kockums, Vestas, Ocean Wind Energy Systems, the U.S. Department of Energy, Sandia National Laboratories and NASA. Mr. Braun is a graduate of Arizona State University and the author of numerous technical papers on renewable energy technologies, and the transition to solar hydrogen production, storage, transportation and end-use systems. Mr. Braun has been an advisory board member of the International Association for Hydrogen Energy (Coral Gables, Florida) since 1981, and he serves as the Director of the Hydrogen Fleet Project. *For more information about Mr. Braun, you may contact* **phoenixproject.net** *or* **sustainablepartners.com**